SMART
TEXTILES

FOR DESIGNERS

INVENTING THE FUTURE OF FABRICS

SMART TEXTILES

FOR DESIGNERS

INVENTING THE FUTURE OF FABRICS

REBECCAH PAILES-FRIEDMAN

Laurence King Publishing

Published in 2016 by
Laurence King Publishing Ltd
361–373 City Road
London EC1V 1LR
e-mail: enquiries@laurenceking.com
www.laurenceking.com

A catalogue record for this book is available
from the British Library.

ISBN: 978-1-78067-732-3

Design: Lizzie B Design
Design of chapter openers: Nicolas Kubail Kalousdian
Picture research: Sophie Hartley
Senior Editor: Peter Jones

The network pattern in the cover design is derived from a
data representation of the First Amendment created by Chris
Alen Sula and is based on an original design by Nicolas
Kubail Kalousdian

Printed in China

Introduction

This book has been an incredible journey for me. I am amazed at the enthusiasm and interest the topics of smart textiles and new materials bring. Never before have so many people been intrigued with the combination of art, design, and science. The e-verse is constantly buzzing with the introduction of each new smart product, futuristic material, and revolutionary manufacturing process. Wearable technology and smart textiles are hot topics with a growing number of people interested in exploring how these new materials can be used in both their artistic practice and personal life.

The revolution has begun. Smart textiles will redefine the way we think of our clothing, homes, and other products. It is now the norm to exercise while monitoring your heart rate, steps, exertion rate, and hydration among other metrics. Once debilitating health issues are now manageable, giving patients mobility, freedom, and transparency to the workings of their bodies, resulting in an improved quality of life. Designers are working in tandem with scientists to develop textiles with multiple benefits ranging from incredibly lightweight, insulating and cooling textiles to textiles that transmit light, sound, and smell. Other developments include textiles that transform their structure and textiles that are created with sustainable production methods including waterless dyeing techniques and biological processes that grow textiles in the lab.

Information gathering from our actions, bodies, and environments has never been so pervasive as it is today and this is only the tip of the iceberg. Words like disruption, privacy, metrics, and augmentation are commonplace. But what exactly are these new smart materials and what is the benefit of wearable technology? Many designers, artists, researchers, scientists, technologists, and engineers are working on creating these new technologies and applying them in their work while others explore their impact on society, the environment, and ourselves. We are only at the beginning of this field of exploration.

Sparked from my personal experience as an industrial designer developing active sports products and apparel, these new smart materials beckoned me like a bright light. I spent over a year immersing myself in the research and exploration of smart textiles and the work being done in the application of these emerging technologies. The result is this book, *Smart Textiles for Designers: Inventing the Future of Fabrics*, an introduction to smart textiles and the materials and processes that are being explored and applied in many labs, studios and workshops around the globe.

Rebeccah Pailes-Friedman
Brooklyn, NY, 2015
www.getinterwoven.com

DEFINITION AND CONTEXT

Truly new ideas are rare but when one comes along it can start a revolution. We are at the frontier of one of these revolutions today, the techno-textile revolution, the convergence of textiles and technology.

The materials that you will be introduced to in this book will challenge your idea of what fabrics and textiles are and inspire you to rethink what your clothing and other products made with textiles can do.

Fabric is our second skin. We each have a deep personal connection to it from the time we are born. Every day we expect the fabrics that touch our bodies to keep us warm or dry, to be soft or strong, to be flexible or give us support and many other things. At the same time, we demand that our products and clothing perform many of these functions while maintaining their shape and color and expect that they will be easy to care for and to clean. We use textiles in our environments and to build with, we use them in manufacturing and require that they be sustainable, renewable, and use little energy. The more advanced textiles become, the more we demand of them. Textile and material science is constantly developing ways to meet, exceed, and even anticipate our demands as consumers and users.

The pioneers of this field explore the next frontier of these technological developments in materials, fibers, fabric construction, and chemical finishes.

Interview: Melissa Coleman

New media artist, lecturer, blogger, and curator of "Pretty Smart Textiles" and three smart textiles exhibitions, Melissa Coleman has her pulse on the future. Her works are critical explorations of the body in relation to technology. She writes for Fashioning Technology and has taught at art and design schools in The Hague, Rotterdam, Tilburg, and Eindhoven. I interviewed Melissa to get her perspective on the difference between smart textiles and e-textiles and what she thinks are the most exciting things that are happening in smart textiles today.

RPF: Can you give me a simple definition of e-textiles and the difference between smart textiles and e-textiles?
MC: E-textiles is a name for textiles that integrate electronics in the textile itself. In its most perfect form, which doesn't really exist yet, you might not even be able to tell the difference between a regular textile and an e-textile because the electronics are physically part of the textile structure. At the moment this is mostly happening in the form of metal-coated textile fibers. But people are also experimenting with fibers that can generate electricity through movement, or through the sun. Smart textiles don't describe the material as much as its innovative character. In some cases this might mean a highly technical nano-coating, in other cases it might mean the textile has integrated electronics and a computational function. Really all the "smart" part describes is that it has more functions than a traditional textile.

RPF: From your perspective, what do you see as the most exciting things happening in smart textiles since you curated "Pretty Smart Textiles"?
MC: Well, there are two works that I think are particularly interesting because of what they suggest in terms of future visions. One work is by Jalila Essaïdi, she's a Dutch artist and curator who founded the BioArt Laboratories in Eindhoven. One of her works, *2.6g 329m/s,* involves integrating spider silk into human skin. Spider silk thread is much stronger than steel and if we were to get such a skin transplant—or if our bodies could produce it—we could potentially resist the impact of a bullet. She created a little patch of human skin with embedded spider silk, and when a bullet is fired it bounces off it, so it's actually a tiny bit of

The Holy Dress, by Melissa Coleman in collaboration with Leonie Smelt, is a garment that trains you to be a better person. Using a speech recognition system and voice stress analysis, the dress starts to glow and increases in intensity as the likelihood of a lie is detected. When it guesses that a lie has been told, it lights up fully and then flickers as it delivers an electric shock to the wearer as punishment for the lie. For the wearer of *The Holy Dress*, technology performs the function of religion, helping the wearer live an honest life.

bulletproof skin. There are still problems in making it work in the long term in a real living body, but I like to think that's just another technical problem to be solved. This project's main function, in my opinion, is not making this work but in showing that it might work. This triggers the imagination. What do we want to become if we can make technology integrate with us on the most intimate levels?

RPF: That sounds really interesting.
MC: It is super interesting. I think, even though it's such a small patch it suggests so much. It just breaks open the idea of what our skin could be once we start manipulating it in all kinds of ways. What's funny is that she never considered it a textile until other people started calling it that. For her it was much more about specific augmentation of the human body. But I think for people who are interested in textiles the two are closely connected. Clothing is sometimes called our second skin. To me this project suggests a merging of our first skin that we were born in, and this second skin that man created.

RPF: And the other work?
MC: This work is by Ebru Kurbak and Irene Posch called *Drapery FM*. It's a material exploration into how you could incorporate electronics in the structure of a textile. I think the original idea that they started with was political where they were trying to make sweaters that would be able to do peer-to-peer communication so that you could exchange data on a highly personal basis.

In the end they made a knitted fabric that functioned as a radio. You could tell the different electronic components (resistors and capacitors) apart because they used different colors of thread that had a metal thread in the twining. The innovative aspect was mostly in creating textile capacitors through knitting. There's a fortunate coincidence where the way the knitted textile is built up through loops sort of matches the way an electronic capacitor works. So they were able to make really seamless electronics. The beauty is in how much it is still textile and how much it works as a piece of electronics. That works on the imagination in a very different way.

INTERVIEW: MELISSA COLEMAN

RPF: All of these artists seem to be working in an area that questions where our bodies end and technology begins. Is this a theme you see fine artists who work with smart textiles exploring?

MC: One of the functions of fine art is to create a discourse around subjects that are already happening in society, or that are looking to start happening. And anything related to textiles is never separated far from issues around the body. When you start combining textiles and electronics it touches on issues around privacy, intimacy, expression, and different forms of display. It makes sense to me that fine artists would be interested in this medium because it's a very suitable medium to discuss these topics.

RPF: I can see that using smart textiles and wearable technology can be a great way to visually show what your body is doing in a way that you wouldn't necessarily be able to otherwise.

MC: Yes, it's a lot about identity and the personal aspects of technology. I think that technology, even though a lot of it's happening online in a global semi-public space, is becoming hugely personal as well. It's that strange thing where technology makes space, time, and social context irrelevant to some degree. The borders between public and private are also blurring significantly. Who knows, it may yet get to the level where we will literally want to share everything about ourselves.

RPF: Which media art organizations in the Netherlands do you recommend following?

MC: *V2_Institute for the Unstable Media* and *Mediamatic* have been instrumental in popularizing e-textiles and wearables in the Netherlands over the last ten years. But it's not just media art organizations anymore. A lot of interesting things are happening in universities and art and design schools in The Netherlands. Eindhoven University of Technology even has a part of their industrial design program especially dedicated to it, called Wearable Senses.

The Second Skin

What makes smart fabrics revolutionary is that they have the ability to do many things that traditional fabrics cannot, including communicate, transform, conduct energy, and grow. Think of the body as a communication device. Our five senses are the input and output tools; they are how we give and receive information about what is happening inside our bodies and outside in our environment. Our garments interact with all of our senses; they are seen, heard, felt, smelled, touched, and may even be tasted. Smart textiles join these phenomena by using our senses as a way of gathering information from and about us by means of pressure, temperature, light, low-voltage current, moisture, and other stimuli.

Some smart textiles also have the ability to gather this information from our bodies and transform it into data, which can then be communicated in a number of ways. A fabric might form a chemical reaction or transmit low-voltage electric pulses for a software-based computation, for example, or fibers, yarns, or an entire fabric might transform in reaction to the information. It is this ability to

1.3
Designer Flori Kryethi 3D prints three continuous ambiguous forms out of recently developed elastic then joins them together to create the *Trip Top* conceptual garment.

1.4
Developed by Leah Buechley at Massachusetts Institute of Technology (MIT)'s Media Lab, this "living wallpaper" uses magnetic and conductive paints in combination with temperature, brightness, and touch sensors to create interactive walls that can turn on lights, control heat, and manage your music.

respond to external stimuli that gives "smart" or "intelligent" textiles their name. Smart textiles "learn" from our bodies and our environments, and react.

Extroverted and introverted

These reactions come in two basic forms—extroverted and introverted.

The extroverted reaction provides an obvious external transformation that is experienced not just by the wearer. For example, clothing lights up, changes shape or color, creates sound, or gives off an aroma.

When it comes to stimulating the senses of an audience, the possibilities for smart textiles include everything from visual effects to sound, touch, smell, or taste. Embedding fabrics with digital devices aided by microprocessors creates

1.5

1.5
The Human Antenna is a project that uses the body's ability to receive radio waves; these are transmitted to a carpet made out of conductive thread in order to create or capture sound. When a person steps onto the carpet, they supply the carpet with the radio waves they receive. The carpet then makes those radio waves audible.

many of these functions. An early example of this is fabric woven with conductive yarns that enable the control panel of your iPod or MP3 player to be manipulated directly through the sleeve of your jacket. As well as belonging to the category of smart textiles, these fabrics are also often referred to as "e-textiles" (see page 27).

In the introverted reaction, the textile reacts to a stimulus but it does not necessarily display a visible physical change. Instead, the fabric transforms so that one or more of the wearer's own senses are stimulated. Fabrics that can react to a wearer's body have been around for many years. For example, a large category of technical fabrics developed primarily for athletes is focused on comfort, heat control, and moisture management. One example is fabrics that are designed to keep athletes dry. As the wearer starts to perspire, these fabrics absorb the moisture and then react by wicking it to the surface, away from the body. Once at the surface, the moisture spreads out, thus speeding up the evaporation process. This reaction achieves two functions: The wicking keeps the layer of fabric that is closest to the skin dry, and the evaporation keeps the wearer cool. Most of these fabrics create this reaction mechanically, through the construction of the yarn or fabric, or with a coating or finishing process.

1.6
Fashion designer Hussein Chalayan uses built-in technology to explore the "concepts of disembodiment and metamorphosis" by creating cutting-edge garments that change form.

1.7
Polychromelab has developed a jacket for mountain sports that uses smart fabric designed to cool when it's warm and to warm when it's cool, making it ideal for the extreme temperature swings experienced while mountaineering.

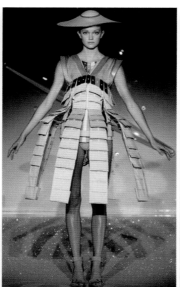

Another moisture-management example is Gore-Tex®, which is produced using a process developed by Wilbert Gore and his son Robert in the late 1960s and early '70s. The Gores applied a membrane to the reverse of a fabric, which allowed smaller air molecules to pass through but blocked the larger water molecules, thereby creating the first breathable, windproof, and waterproof fabric.

The latest generation

When they were first introduced, both of these technologies were cutting-edge developments, but they have now been around for a long time and are readily available on a multitude of products. The latest generation of smart fabrics with moisture management have the ability to sense your body temperature and react by transforming their molecular structure to generate, retain, or give off heat as required to keep your body temperature comfortable.

Innovations in nanotechnology—the science of manipulating atoms, molecules, and materials on the scale of nanometers—have also yielded super-absorbent materials, finishes, and coatings that work to manage moisture faster, and absorb more moisture than any previous material.

1.8
CuteCircuit partnered with Mercedes to create the Pilot Suit, a driver's jacket of 16,000 white pixel lights controlled by biosensors. The garment emits light according to the wearer's feelings and emotions while they drive.

Smart Materials

Smart textiles are the textile version of smart materials. Traditional textiles are made from yarns that use materials chosen for their mechanical or structural qualities. Silk for its strength, light weight, and its affinity to dye; wool for its insulating ability; and polymers for their stretch, comfort, and price. The introduction of smart materials into textiles brings their inherent qualities to a flexible, wearable, and easily manufacturable product.

Smart materials have been around for years. The term "smart" or "intelligent" was first introduced in the US in the 1980s even though many smart materials had been in use for many years before then, but the introduction of smart materials to textiles is relatively new.

There are three categories of smart materials based on their functions; passive, active, and very smart materials. Each of these levels involves different types of technology. The lowest level of function is passive smart materials. They act as sensors, sensing the environment or stimuli. They gather information and can show what is happening on them such as color change, thermal or electrical resistivity. For example, a fabric that changes color when your body temperature changes. Photochromatic inks are pre-programmed to trigger at a particular temperature and to change their hue. Exposure to UV light waves creates the reaction.

1.9

1.9
Using photoluminescent fabric, Grado Zero Espace and Bertone collaborated with Alfa Romeo to create futuristic concept car seats that are self-illuminating.

The next level of smart materials is active smart materials. These materials have the ability both to sense and to respond to external stimuli. When they are exposed to an environment, they act as both sensors and actuators. A number of active smart materials generate voltage when they are exposed to pressure, vibration, changes in pH, a magnetic field, or temperature. For example, when applying stress to a piezoelectric material voltage is created. A piezoelectric material is one that releases the same charge that is put into it. Because the reverse of this reaction is also true, when a voltage is applied to the material stress is created. This reaction has led to the development of materials that bend, expand, and contract when electrical current is applied.

Finally, very smart materials add a third function to the equation. These materials act as sensors and receive stimuli; they can react to information; and they can reshape themselves and adapt to environmental conditions. This category of materials is one of the most advanced and dynamic areas of research and discovery leading to new and exciting products and product categories: It includes shape-memory alloys, smart polymers, smart fluids, and other smart composites.

1.10
Designer Angella Mackey, working with Toronto's Social Body Lab, has developed a wearable light that is designed to blend with your existing clothing. Crafted from LEDs controlled by a small circuit board, the Vega Edge can connect to the edge of any garment or accessory.

1.11
The Wall E-(motion) by Diffus is a series of applied discs made out of textiles and conductive thread. The discs can be arranged to generate different sensations including lights, scents, and sounds.

1 DEFINITION AND CONTEXT

Emotive Textiles

Emotive textiles trigger your senses using color, light, scent, or sound. These effects are achieved through chemical compounds embedded in the fiber or yarn, or used to coat the finished fabric. Nanotechnology also facilitates advanced dyeing and finishing processes, which can expand the properties given to a fabric; the finished result is more powerful and longer lasting. Research is being carried out in many universities into nano-coatings, inks and dyes that change color, and ways of embedding scent and other properties into textiles, while the US military is working on printed uniforms that change color in response to their surroundings, rendering traditional camouflage obsolete. Once developed, these technologies will have far-reaching applications.

1.12 & 1.13
Using a non-invasive EEG headset to measure brain wave activity while an individual listens to music, the NeuroKnitting project translates the subject's reactions into knitted fabric. Artists Varvara Guljajeva and Mar Canet collaborated with MTG researcher Sebastián Mealla to produce these custom scarves, using their open-source "Knitic" knitting machine. Each unique scarf is a reflection of each person's brain activity at a single point in time.

1 DEFINITION AND CONTEXT

One of these applications is illustrated in *Herself*, a couture textile sculpture that is the first dress able to purify the air around it. First revealed in Sheffield in the UK in 2010, *Herself* is an artistic interpretation of the idea that textiles may be able to eliminate certain pollutants from the air so that we can breathe more easily.

Another growing field is that of micro-encapsulation, which involves the use of coated "microcapsules" to provide materials and products with particular qualities. The technology has a number of applications—for example, it is used in antimicrobial deodorants, sunscreen, and time-release pharmaceuticals. In the world of textiles, micro-encapsulation technology can be used to bind fragrance to individual fibers, which are then spun into yarns, and woven or knitted into fabrics.

1.15

1.14
Helen Storey collaborated with three universities in the UK to create the conceptual *Herself* dress as part of the larger *Catalytic Clothing* project. The project seeks to engage the public in the science of controlling environmental pollution through textiles.

1.15
Inspired by the form of the human lungs, the dress uses smart textile technology to improve ambient air quality by eliminating pollutants. Although the dress is a conceptual art piece, it is underpinned by sound science.

Designed for Protection

Smart textiles have also been evolved to react to their environment without electrical input. Using materials science, molecular technology, and nanotechnology, these textiles are engineered to act as an outer shell for our bodies, protecting us from harm and allowing us to achieve greater feats. They increase our ability to withstand extreme temperature, fight off attack, and protect us from gunfire. They can administer medications, monitor our vital signs, heighten our awareness, elevate or monitor our mood, emit scents, regulate our body temperature, and keep us dry. Fueling many of these discoveries is scientific research funded in part by governments for space or military advancement.

Military research has led the quest for new and improved developments in body armor and the increased protective qualities of soldiers' uniforms. Uniforms are now not only more protective; they are also lighter, more comfortable, and more durable, thanks to new fibers and nanotechnologies embedded in their fabrics. Uniforms are now designed to keep a soldier's body at a constant temperature, while protecting him from the impact of bullets, and providing communication between the field and command. Projects are even underway to develop uniforms that can apply pressure to a wound and administer medication to stop bleeding while transmitting vital signs to a command base.

1.16
Benetton team driver Jos Verstappen escapes a race car accident with only minimal burns. His life-saving racing suit is made with DuPont's fireproof Nomex® fabric.

1.17
Specially designed cool suits and helmets protect the McLaren pit crew from the intense Formula One trackside race environment. The suit uses a cooling pump to circulate liquid through small tubes embedded in the fabric, keeping the crew member's body temperature manageable. A full face helmet is lined in Nomex® and has a flameproof retractable visor.

1.16

1.17

The cutting-edge research behind these developments is hitting the street as well as the battlefield, as researchers and entrepreneurs apply these discoveries to new products. For example, Ministry of Supply—a venture set up by a team of recent graduates from MIT—is producing products that combine the performance of technical bicycling apparel with the look of everyday business clothing.

To help understand how the body's skin moves, the team turned to an engineering process used in aerospace design. Using thermal imaging to find the spots in the body that generate the most heat, they designed their dress shirts with strategically placed areas of stretch, and vents to move with the body and regulate temperature.

In addition to mechanical design details, Ministry of Supply licensed a fabric that NASA had developed to regulate astronauts' body temperatures in 390°F (200°C) heat. Adapting it for their clothing, this fabric keeps the body cool, preventing perspiration from occurring in the first place.

The work of this company provides an excellent example of how advanced research now reaches our everyday lives.

1.18
DuPont's Kevlar® aramid fiber has been used in helmets, vests, and vehicle armor for military and law-enforcement personnel to protect from bullets for over 35 years. New advances have made Kevlar® XP lighter weight and stronger, improving safety and increasing productivity.

1.19
MIT professor of aeronautics, Dava Newman, has developed the BioSuit, a spacesuit that uses a combination of skintight smart fabrics to apply "mechanical counterpressure" equal to 30 percent of atmospheric pressure, the magic number needed to keep someone alive in the vacuum of space. The revolutionary suit is designed to support life in atmospheres like that of Mars while increasing the comfort, mobility, and flexibility of the astronaut.

1.18

1.19

E-textiles

Innovations can originate in unusual ways, and some of the world's greatest have been created by mixing dissimilar ideas together. It took a visionary to think of combining computing and textiles. The results are fabrics that can transform, collect, and transmit data, store and conduct energy, and house computers that are lightweight, flexible, incredibly small, and washable. Just a few years ago such fabrics would have seemed like science fiction. After all, conventional computers are rigid, heavy, large, and run on electricity, and their most common interfaces are a keyboard, mouse, or trackpad. Fabrics, on the other hand, are soft, sensuous, lightweight, and flexible. They are usually not very strong, are often exposed to getting wet, and are not thought to be capable of conducting electricity. In other words, the two fields could not be more different.

However, in the mid-1990s a team of researchers, led by Maggie Orth and Rehmi Post, at MIT's Media Lab began to explore how digital electronics could be integrated into clothing by using conductive materials that could be applied, sewn, embroidered, or otherwise incorporated into fabrics. Their work was the birth of e-textiles, and the seed that started the techno-textile revolution. Orth's pioneering works of art experimented with conductive fibers embedded into yarns colored with thermochromatic inks and woven into fabrics that then changed color with the introduction of an electric current (see page 134).

1.21

1.20
The talented husband-and-wife design team of Pankaj and Nidhi from India have created award-winning gowns, dressing a global clientele in their amazing creations. Here is one of their glowing geometric dresses from the Spring/Summer 2012 collection, shown at Wills Lifestyle India Fashion Week.

1.21
The futuristic Geometrica collection was filled with intricate latticework jackets, origami-inspired embroidery, geometric forms, and four amazing light-up Glow dresses. Pankaj and Nidhi utilized emerging garment technologies of laser cutting, digital printing, and wearable electronics to create some of the most experimental fashions today. The Geometrica collection won accolades as the most ingenious collection shown at Wills Lifestyle India Fashion Week, Spring/Summer 2012.

From science fiction to reality

Since then, developments in e-textiles and wearable technology have grown exponentially. As information technology has become more integrated into our everyday lives, wearable computing has moved from science fiction to reality, as researchers and manufacturers continue to bridge the gap between electronics and textiles.

As the lines between computing and textiles blur, so do the lines between computing and our bodies. Wearable technology is in the process of evolving to a point where it will pervade our lives; we will soon become very accustomed to being surrounded with technological conveniences. Our garments will become mobile devices that seamlessly connect us to our networks through interfaces that will be so fully integrated into our actions and activities that they will be more pervasive than invasive.

Textiles are the bridge between something we are all very familiar with, and ideas that are so sophisticated that only a few advanced research engineers and scientists can understand. The fields of smart textiles and wearable technology are completely intertwined. With e-textiles, information is transmitted through fabrics woven from conductive yarns. The fibers used to compose these yarns are very flexible and easy to manipulate, and consist of a combination of materials—including carbon for strength and abrasion resistance, polymers, and finely drawn copper, coated with metals like silver and nickel (which are excellent conductors). Very small silicon chips and sensors have also been developed that can be embedded with the fibers as they are spun into yarns. These are then integrated into tiny flexible circuit boards.

Such devices have been engineered to feel and perform like a regular textile: They have drape, flexibility, and can be made to a specified thickness, surface texture, weight, and durability. Once woven and sewn into clothing, they provide a garment that a general consumer can respond to with much greater ease than with a traditional electronic device. The benefits of bringing technology "out of the box" will revolutionize how we approach many aspects of our current lives, perhaps even creating the next industrial revolution.

1.22
The German company Novanex, specializes in integrating wearable technologies into textile products. Working with Fraunhofer IZM and Stretchable Circuits, Novanex has created an interactive system toolbox for costume designers to make it easier to create individually programmed light effects on their costumes based on integrated sensors.

Beyond Woven

Beyond weaving and knitting, new manufacturing methods are helping create wearable materials that are not strictly textiles in the traditional sense. Designers are starting to use 3D printing in polymer-based materials to create wearable garments and accessories. 3D printing creates a three-dimensional object from a digital file through an additive process where the printer lays down successive layers of material until the object is created. 3D printers are capable of printing many different materials from plastics, paper, ceramic, glass and metals. Although this process is still in its infancy, 3D printing will revolutionize the way we create and acquire products. The application of this manufacturing technique to wearable products is an exciting and evolving area of study.

1.23
This 3D dress for Dita Von Teese designed by the Francis Bitonti Studio in collaboration with Michael Schmidt Studios is made from 17 individually printed nylon pieces that are similar to traditional pattern pieces. Each of these pieces is joined together and then decorated with Swarovski crystals to create a fully articulated gown.

1.23

Additionally, there are numerous projects that explore the emerging field of wearable technology. Although not strictly a smart textile, the projects that attach technology to the body are the first attempts to utilize movement, temperature, pH, and other factors in a product we wear. Soon textiles will become the actual technology and the incorporation of electronics onto the surface of a textile will be forever transformed. Smart textiles are not limited to practical applications. Artists and designers are inspired by the intersection of technology and nature. A number of individuals are creating work that challenges our understanding of beauty by using non-traditional materials. Many fashion designers use traditional fabric combined with smart materials and wearable technology to create modern, thought-provoking and aesthetically beautiful pieces.

Whether smart textiles are wired to transmit data or have technology embedded within their construction, what sets them apart from traditional fabrics is that they have the ability to adapt or react to stimuli. The focus of e-textile development is on the seamless integration of electronics into textiles. From haute couture to artistic performances, medical devices, safety gear, first responders' equipment, and extreme-condition applications, smart textiles have evolved to protect, enhance, strengthen, entertain, and amaze us. In Chapter 3, we will explore how these new materials are being used in a survey of projects by researchers, designers, and manufacturers across disciplines and from around the globe.

1.24
Fashion designer Mathieu Mirano creates a magnificent gown with an iridescent bodice made from real beetle wings. His unique textures and colors are captured with unusual combinations of traditional and non-traditional textiles.

1.24

1.25
Sound artist Di Mainstone crafted a
wearable musical instrument, the Human
Harp, to harvest the vibrations of the
Brooklyn Bridge's suspension cables.
The "parasitic instrument" attaches
to the bridge magnetically and uses
digital sensors that detect and measure
the vibrations of the cables as the
wearer moves.

By extending, plucking, and moving
with the Human Harp's retractable
musical strings, the user can adjust
various characteristics of the structure's
voice, uniting movement and
music. Human Harp is now a global
collaboration connecting engineers,
dancers, musicians, and bridge lovers
from around the world.

1.25

2

MATERIALS

Technology is not new to textiles. The exploration of new materials and material combinations, alloys and blended yarns, and the manipulation of the basic textile components, including new fibers, yarn shapes, and textile construction, has evolved to become one of the most exciting fields of science and technology today.

The exploration of smart textiles has led to new research in protective clothing, performance athletics, fine art, fashion, product design, architecture, healthcare, and communications.

Materials scientists, engineers, and designers have worked together with athletes, first responders, and the military to develop fabrics that have increased speed, endurance, and protection. Fabrics have been developed for protective clothing that has saved lives from environmental hazards ranging from fire, to radiation and chemicals, and been used in outer space.

Today's athletic apparel fabrics can manage energy output, control body temperature, monitor heart rate and other physiological reactions, and even transmit body position and movement to a virtual coach. Beyond the playing field, there are developments in the field of health and beauty with medical textiles that release drugs and monitor the immune system, there is smart-wound dressing, and a huge range of cosmeto-textiles developed around aromatherapy, skin treatments, and perfumes.

This chapter offers a survey of smart textiles, along with some of the tehnology behind their development.

Basic Textile Components

There are six components in textile materials: **fibers**, **yarns**, **fabric construction**, **fabric interstices**, **colorants**, and **chemicals**. The combination and manipulation of these components is the foundation of textile science, and where we start building our reference library of design knowledge.

Fiber is the term used to describe the needlelike shapes that form the basic unit of any textile material or yarn. Fibers are usually small and vary in length. There are many different types, and they are generally categorized into two basic groups, natural and man-made.

Natural fibers are naturally occurring and derived from animal, plant, and mineral sources. Animal fibers are generally protein fibers, and include animal hairs such as wool, alpaca, cashmere, mohair, vicuna, angora, as well as silk (from the cocoons spun by moths). Plant-based, or cellulosic, fibers are sourced from plants, trees, and flowers, and include cotton, flax, ramie, jute, hemp, and coir. Mineral fibers are mined from the earth and include asbestos, copper, and silver.

Man-made fibers are manufactured from a material that at any point in the process of being manufactured was not a fiber. Man-made fibers can be categorized into three groups: They are either a transformation of a natural polymer (acetate, cupro, modal, rayon, rubber, or viscose), a synthetic polymer (acrylic, aramid, nylon, polyester, polypropylene, or spandex/elastane), or they are inorganic (carbon, ceramic, glass, or metal). There are a great number of man-made fibers, with a wide range of performance properties. Research is constantly yielding new and exciting fibers and fiber combinations. Many of the resultant new materials will be introduced in this chapter.

2.1

Yarns are continuous strands of twisted fibers, used to created a textile by weaving, knitting, or other forms of intertwining. Yarns can vary in size, strength, diameter, uniformity, and makeup.

Fabric construction refers to the arrangement of yarns—or sometimes the fibers themselves—within a textile. In weaving, yarns are interlaced at right angles to each other. There are many variations of woven patterns, from a basic weave to complicated Jacquards and double-woven patterns. Woven fabrics are usually stable and do not have much stretch (unless the yarns themselves have stretch properties).

Knitting uses a single yarn that is continuously interlooped. In general, knits have more give than woven textiles, but not always. The pattern in which the yarns are looped will be a key indicator of the finished textile's characteristics.

Entangling involves creating a textile directly from fibers without creating a yarn. Felt and Tyvek® are examples of entangled textiles. The fibers are held together by friction between the fibers.

When working with textiles it is important to consider the **fabric interstices**, or the spaces between the yarns or fibers. This non-fiber area makes textiles extremely porous—60–90 percent air by volume—leaving a much smaller percentage of the volume in "solid" material. This porosity adds to the

comfort of textiles and can also lead to other advantages: The interstices may be filled with resins or other chemicals that impart properties ranging from waterproofing to UV protection, moisturizing effects, and antibacterial action. Air trapped in the interstices also contributes to a textile's insulating properties, while the porosity makes them breathable.

There are numerous ways that textiles can be finished, including adding colorants and using chemical treatments to create different surfaces and physical properties. Colorants, including dye stuffs and pigments, can be added at the fiber, yarn, or finished-textile stage. These chemicals adhere to the fibers and become part of the physical makeup of the textile. Colorants also add weight and can change the texture and hand feel of the finished fabric, as well as its color.

Other chemical treatments can be used according to the desired application, such as UV protection, flame retardation, water repellency or waterproofing, stain repellency, perfume, and many others. Chemical finishes may be applied directly to the fibers, embedded within them, as a film around the yarn or fiber, or applied within the interstices by spraying, printing, or dipping the finished goods in a chemical bath. Chemicals are also used as a coating or layer on one side of a fabric to create a barrier, or to bond two pieces of fabric together.

The Birth of Performance Fabrics

Fabrics with performance characteristics are not a new phenomenon. Natural fibers have long been valued for their comfort and ability to help regulate body temperature in both heat and cold. For example, yarns made from sheep's wool can be considered among the first performance fibers.

Wool has a unique ability to keep a wearer warm even when it's wet; it's hydrophilic, since it can absorb more than 29 percent of its weight in moisture; and it's also hydrophobic—it will not immediately absorb the moisture from rain and snow but instead will repel it, making it an ideal fabric for outerwear. It's naturally flame retardant, too. Although bone-dry wool will ignite, most wool retains enough moisture that it will self-extinguish. And because wool is a protein fiber, it reacts well to organic solvents and stain removers and can be cleaned without being damaged. Wool is exceptionally resilient, too, and recovers from crushing and creasing, yet will allow pleats and durable creases to be made through the application of steam and pressure— the combination of heat, moisture, and pressure work together to deform its original molecular structure to hold the desired crease in place. All these performance characteristics are inherent to a fabric that has been in use since before recorded history, and wool remains an inspiration to textile scientists to this day.

The molecular science of natural fibers, like wool, serves as a resource for textile scientists and designers in the study and creation of engineered fibers. Scientists study the properties of natural fibers and strive to recreate these characteristics in synthetic materials to better simulate or even enhance what nature does on its own. For example, Kevlar®, a synthetic fiber developed by DuPont in 1966, is an aramid fiber characterized by its high strength, and best known for its use

2.1
Chemical finishes, such as the nano-coating of this super-hydrophobic fabric, can be utilized for such things as UV protection, flame retardation, and water repellency, as well as many other uses. They may be applied directly to the fibers, embedded within them, as a film around the yarn or fiber, or through spraying, printing, or dipping in a chemical bath.

2.2
Designed with an equal blend of form and function, the Italian-made Spidi kevlar jeans are a cotton/Lycra/Kevlar® blend. They resemble ordinary jeans on the surface, but are reinforced with Keramide synthetic fibers underneath for protection from abrasion.

2.2

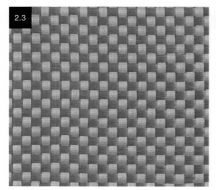

in body armor. It was a revolutionary discovery and has saved countless lives, but to be effective it must be completely waterproofed via a lengthy, expensive process. Recently, in an Australian study, scientists discovered that adding tightly woven wool to Kevlar® improves the energy and water absorption of the final material and enables the creation of bulletproof vests that are lighter in weight, and less expensive to produce. The researchers also discovered that the wool expands when it absorbs water, strengthening the vest's overall penetration resistance and reducing the number of layers of Kevlar needed to stop a bullet from 36 to 30. In another example, wool is being re-engineered to enhance some of its natural characteristics while altering some of its less desirable tendencies, like shrinkage. Total Easy Care (TEC) Merino technology is a permanent treatment that is either applied to the spun yarn or to the garment during the finishing process. The process consists of an oxidative pre-treatment followed by the application of a polymer that swells during washing and seals the scale edge of the wool fibers that create the felting effect during normal washing. Garments made from TEC Merino are all machine washable, and some can even be tumble dried.

Designers and scientists have long sought to manipulate the basic components of a textile to achieve desired effects, and many of the early iterations of performance fabrics predate this current smart-textile

revolution. The field of performance textiles was born around the time of the 1912 Olympic Games when, for the first time, competitors came from all five continents to compete in Stockholm, Sweden. As competitive athletics became more popular and the interest in developing specialized clothing increased, companies started to develop fabrics that could improve athletes' comfort, movement, and performance.

Pique fabric

In the late 1920s, French tennis player, René Lacoste, the seven-times Grand Slam tennis champion, worked with designers to create an entirely new fabric, which he wore for the first time in the 1926 U.S National Championships (now the US Open). His creation, the pique knit, has long been considered the original performance fabric. The pique knit is a 100 percent cotton fabric that

2.3

Best known for its high strength and use in body armor, Kevlar® is a synthetic aramid fiber developed by DuPont in 1966 that has saved countless lives, despite its lengthy and expensive waterproofing process.

2.4

The synthetic fiber Kevlar® combined with tightly woven wool improves energy and water absorption, contributing to lighter weight and lower costs. When it absorbs water, the wool expands, thus strengthening the armor's overall penetration resistance.

has a textured face with many small mesh-like holes and a smooth back. This construction creates a much larger surface area on the face of the fabric relative to the back, enabling moisture from the wearer's skin to be transported from the back of the fabric to the face and to spread over the larger area, thus speeding up evaporation. The evaporative process, in turn, cools the athlete, thereby performing double duty: the wearer remains cool and dry.

Pique fabric is an example of a fabric innovation that was designed as user-specific athletic wear. The combination of the absorbent cotton fiber and the fabric's unique construction created performance characteristics that have endured for almost a century. This same construction has inspired thousands of variations and countless combinations of fibers and yarn weights since the 1920s, and many of today's performance fabrics still use variations of the original pique construction.

The progression from performance fabrics to smart textiles has evolved rapidly, with developments in the fields of nanotechnology and phase change. Led by government contracts and research institutions for the military and aerospace industries, many groundbreaking discoveries have been made. In the past 20 years, research has exploded in directions that we could never have imagined. Today fabrics are engineered for health and wellness, with embedded properties to fight infection, dispense medication, relieve stress, and monitor vital signs, to name a few.

2.5

Micro-encapsulation is used to imbibe a chemical directly into the physical construction of a fiber. Unlike a coating where the chemical sits on top of the fiber, in this technique the desirable properties become part of the fiber. Phase-change materials are incorporated directly into a textile's fibers as seen in this microscopic view from Outlast®.

Survey of Smart Textiles

1: Performance Enhancement

The progression from technical performance fabrics to smart textiles evolved as greater demands were made on fabrics—for example, by those working in extreme environments and needing protection from hazardous conditions; by elite-level athletes pushing themselves to even higher levels of competition; or to keep pace with medical breakthroughs. These textiles now include fabrics with moisture management and other thermoregulators, elastomeric fabrics, and fabrics designed to increase competitive performance by increasing speed, mobility, and endurance. Many of these fabrics can be characterized as first-level smart textiles in that they act as sensors to their environment or stimuli with the ability to react. Some more advanced materials are now being created that both react and transform, but the majority of textiles for increased physical performance are passive smart textiles.

Increased mobility

Perhaps the most well-known performance fibers for increased mobility are those with stretch. Elastomeric fibers can stretch to extreme lengths and then fully recover their original shape, and are used in specialized applications where high elasticity is necessary. This group includes natural and synthetic rubbers, spandex fiber (polyurethane), anidex fiber (polyacrylate), and the biconstituent fiber, nylon spandex (elastane).

The most common use of elastomeric fibers is in conjunction with other fibers when blending yarns to create stretch and aid in recovery. These fibers also have very good insulating properties and resistance to acids. Elastomeric fibers have revolutionized athletic clothing, and have applications in almost every aspect of apparel today.

Thermoregulators

Thermo-regulating fabrics offer a competitive advantage to athletes and others participating in endurance activities or working in excessively hot environments. There are a number of textiles that work with the body to

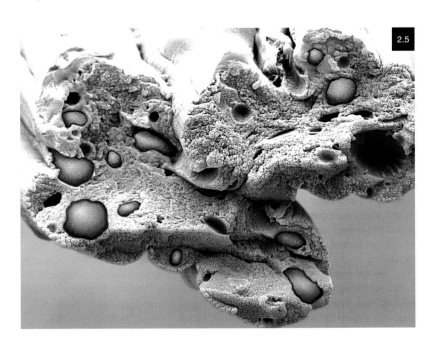
2.5

control energy output and regulate body heat to improve performance.

The human body works most efficiently at a constant temperature. For athletes, it takes energy to "warm up." Then, once muscles are working at peak efficiency and optimal heat is reached, the body expends additional energy to reduce the buildup of excess body heat by sweating and increasing blood flow to the skin. Thermo-regulating fabrics can aid in the capture and release of this heat and energy.

Phase change is the process of a material changing from a solid state to a liquid state, like ice changing to water. The energy gained or lost during this process is the key to the material's ability to "store" or "give back" energy. Originally, NASA developed the technology to embed phase-change materials into fibers for use on astronauts' spacesuits.

Phase-change materials (PCM) are now being incorporated into textiles to create an energy-regulating system designed to keep you warm when you're cool and cool when you're warm, reducing your energy output at both extremes of the equation. These micro-encapsulated materials are embedded into fibers in a proprietary process owned by Outlast® Technologies. They can also be applied to fibers, yarns, or finished fabric as a coating.

Outlast fabric absorbs excess heat from your body and "stores" it as the phase-change materials change over to a liquid state. When the body begins to cool down, the materials begin to solidify, giving off heat, and "return" it to your body. The use of PCM improves comfort by proactively absorbing and storing heat to reduce overheating and perspiration. The overall effect is a conservation of metabolic energy because less energy is focused on skin temperature regulation.

Thermo regulation can also be achieved through the reflection of far infrared rays. Energear™ is an energy-recovery system for textiles developed

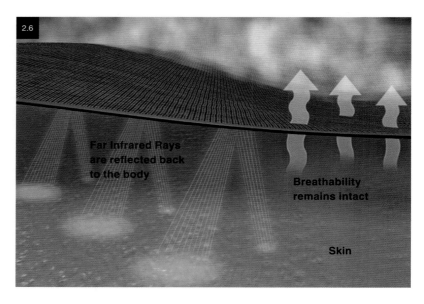

2.6

Far Infrared Rays are reflected back to the body

Breathability remains intact

Skin

by the innovative Swiss textile company Schoeller. This fabric works with a titanium and mineral matrix that reflects energy radiated by the body back to the wearer in the form of far infrared rays. The reflection process promotes blood circulation and the increase of oxygen levels in the blood; this has been shown to enhance performance, prevent premature fatigue, increase concentration, and improve regeneration. In testing, the use of energear™ has also been shown to reduce lactic acid in the blood, and enhance physical performance. (An increase of lactic acid in the blood during exercise leads to more rapid muscle exhaustion.) In stress testing, subjects' heart rates were monitored with and without energear™ clothing; it was found that with physical activity, the heart rate of the subjects with the energear™ outfit could be positively influenced, and in roughly 25 percent of the cases this led to a reduction of up to 20 beats per minute in heart rate—a significant reduction in energy output.

Yet another method of energy regulation is Celliant®, by Hologenix, LLC, which works to recycle captured energy back to the body. It uses 13 microscopic thermo-reactive minerals, which are permanently embedded into natural and synthetic fibers. Over

60 percent of the energy the body consumes is lost through escaping body heat. The minerals work with the body's own energy, modifying the spectrum of invisible and visible light, recycling the energy back into the body by capturing escaping body heat and converting it into infrared energy. Infrared light—not to be confused with UV light—is recognized as having positive effects on the body. It's a medically proven vasodilator, increases tissue oxygen levels, enhances cell vitality, and regulates body temperature. The fiber is used in apparel, bedding, healthcare, and veterinary products.

Coldblack™

Another Schoeller technology that works to regulate surface temperature of the skin and protect from both UVA and UVB rays is coldblack™. Light-colored textiles reflect both visible and invisible rays of sunlight, meaning both light and heat. Dark colors absorb both, and can become much more uncomfortable to wear in sunny conditions. Schoeller has therefore developed a chemical finish that offers full-spectrum UV protection. It can be applied to any textile without affecting its look or feel. It has the effect of reflecting UV off the surface of a fabric with a minimum UPF of 30. When applied to black and other dark colors it lowers the surface temperature of skin and reduces perspiration levels.

Looking to the future of thermo-regulation research, fabrics may soon be able to self-regulate temperature by generating heat on their own to maintain a constant temperature. A team of scientists and engineers from the University of Pittsburgh and Harvard University have worked together to develop a mechanical and chemical feedback loop with synthetic materials that can mimic the human body's ability to regulate its own temperature. Their Self-regulated Mechano-chemical Adaptively Reconfigurable Tunable System (SMARTS) uses tiny "hairs" embedded in a hydrogel. These hairs can stand up or lie down, just as hairs on your arm respond to heat and cold.

SMARTS starts working when the tips of the hairs, standing in the upright position, interact with reagents in an upper fluid layer and generate heat. This heat causes the temperature-sensitive gel to shrink, releasing the

hairs, which are then free to bend away from the reagents again. The system eventually cools down, causing the gel to expand and forcing the hairs back to their upright position to start the cycle over again. This cycle acts as a self-regulating on/off switch. The future of this research will lead to self-regulating microscopic materials that can be applied to everything from energy-saving smarter building materials to biomedical engineering devices and self-regulating smart textiles. This future fabric will actually produce heat when it senses that your body or its environment is cool.

Aerodynamics

In the quest to break world records for speed, athletes on land and water have looked to fabric to reduce drag and create better airflow around their bodies in the hope of shaving tenths of seconds off their times. An example is Speedo's LZR Racer Suit, developed in 2008 and then banned in 2009 when it was discovered that 94 percent of the Olympic swimmers

2.6

Energear™ is a fabric designed to aid in energy recovery. It is constructed to reflect energy radiated from the body back to the wearer by harnessing infrared rays. This diagram shown how energy is reflected back, but allows heat to escape.

2.7

This thermal image of a swimmer illustrates water flow around a swimmer in the pool. Speedo applied this type of information when analyzing water resistance and drag when designing the fabric and seam placement for their racing suits for competitive swimmers.

2.8

Speedo's LZR Racer Suit was so successful in reducing skin vibration and muscle oscillation, as well as decreasing symptoms of fatigue, that it was banned in 2009 due to the competitive edge that it gave. However, advanced swimsuits are still being developed, like the TYR AP12, which implements welding over sewing to improve water resistance.

2.9

2.10

winning medals had worn the LZR suit during competition. The first LZR suit was so startlingly successful that FINA, the International Swimming Federation, were forced to change the guidelines of competitive swimwear in 2010. Co-designed with NASA, the suit's fabric works with compression to reduce skin vibration and muscle oscillation. The suit also has a corset built into the torso to give stability to the swimmer's core. A swimmer's form begins to fail with the onset of fatigue, with the hips dropping and legs dragging. The support gained from the corset gives the swimmer consistent posture during competition, thus improving performance.

Rubberizing, waterproofing, and using neoprene for the surface of materials have all also been experimented with to increase the flow of water over fabric. Speedo's current innovation is the integrated FastSkin3 Racing System, consisting of a swimsuit, swim cap, and goggles. The three elements work together to reduce passive drag, which impedes

the swimmer's glide as water flows over the body. The reduction is achieved with a combination of highly compressive fabrics, body shaping with advanced patternmaking, and no-sew technology in the garment construction. The system is unique in that the elements not only have individual features and benefits, but they also operate together to improve overall performance. Advanced elastomeric fabrics with various surface textures were used to create the optimal flow of water around the athlete as s/he glides through the water.

AeroSwift technology

On land, Nike is the king of innovation on the track. First introduced in 2008, the Nike bodysuit for track and field athletes features a fabric with a surface texture inspired by the dimples on a golf ball. Science has proven that golf balls travel farther and faster when they have a dimpled surface, so Nike designers applied the same thinking to reduce aerodynamic

drag on the athlete. Their AeroSwift technology introduced in the Nike Pro TurboSpeed suit uses a combination of revolutionary patterns and surface architecture placed at strategic points on the body to enhance performance on the track. The bodysuit's technology is not limited to the fabric; it also relies on an innovative, uniform garment construction, which reduces vibration and bulk with a unique, low-profile smooth waistband arrangement, glued and flat seaming techniques, and elastic and edge finishes placed on the outside of the garment. With over 1,000 hours of wind tunnel testing on some of the world's most elite track athletes, the Nike Pro TurboSpeed is estimated to cut 0.023 seconds from the 100m time of Nike's previous uniform.

Downhill skiing and ski jumping are also sports that have relied on advanced design techniques and technical textiles to gain a competitive advantage. Most World Cup competition alpine ski jumping suits are made from a

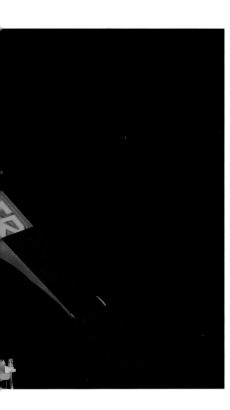

2.9
AeroSwift technology combines strategically placed 3D printed patterns and surface architecture to give runners unparalleled aerodynamic drag reduction.

2.10
Alpine ski jumpers rely on their suits to increase float time, stabilize their movement in the air, and keep them comfortable on the mountain. Technical fabric company Eschler developed a group of abrasion-resistant fabrics from a combination of polyamide/elastane, polyester, and foam that improve moisture management, aerodynamics, and stretch.

fabric produced by the Swiss textile company Schoeller—a ½in- (1cm-) thick multilayered specialty textile with aerodynamic qualities. The International Ski Federation has strict guidelines over the fabric and construction of alpine ski jumping suits, since a suit's fit and fabric can greatly affect an athlete's jump. A looser suit has extra surface area to catch more wind during flight, and allows greater mobility in the legs and upper body. Specializing in warp knits and a proprietary three-layer Eschler Comfort System, the company unveiled a new product for alpine skiers at the 2014 Winter Olympics. They have developed a fabric that merges a polyamide/elastane outer layer and a polyester inside, with foam sandwiched between. It is designed to meet strict new guidelines requiring a more closely fitting alpine ski jumping suit. To add strength and resistance to the fabric's mechanical wear, a coating of diamond-hard ceramic particles is added to the outside. The result is a breathable, abrasion-resistant fabric that combines stretch with outstanding moisture management and aerodynamics.

Another highly anticipated launch at the 2014 Winter Olympics was the Mach 39 speedskating suit developed by Under Armour in collaboration with Lockheed Martin, the giant aerospace contractor. The two companies, both based in Maryland, worked in complete secrecy at the Under Armour Innovation Lab to develop the suit. Mannequins were built to mimic the positions skaters assume while circumventing the track, and over 100 fabrics were wind tunnel-tested in 250 different prototype configurations, while adjusting seams, zippers, and fabric combinations to find the most efficient combination of fabric and construction. The final racing suit uses five different fabrics with a vent running down the spine to release body heat. In line with earlier research done for swimmers, it was discovered that rougher fabrics actually performed better in the wind tunnel and on the

ice by distributing airflow around the skater more evenly. Smooth friction-reducing fabrics are placed in areas where body parts rub—inner thighs and underarms—and rougher surface-texture fabrics are used on the perimeter of the body to optimize airflow.

Moisture management

Moisture management is one of the key performance criteria determining a fabric's comfort, and is defined as the controlled movement of water vapor and liquid water in the form of perspiration from the skin's surface through a fabric. To be comfortable, a fabric must allow the body to maintain a constant heat balance during various activities, and in a wide range of environmental conditions. Breathability refers to the ease with which gases can pass throughout a fabric, and this includes water vapor.

Water vapor is the primary carrier of moisture (perspiration) and excess body heat away from the skin's surface to evaporate, and the evaporative process is key in regulating body temperature. This is not just relevant for comfort in the heat; in cold weather conditions it is critical that the body remains dry in order to maintain a constant temperature.

Since the introduction of the pioneering Gore-Tex® almost 50 years ago, the ability to control moisture buildup in clothing has been a major area of innovation. Today, a new breed of nano-scale finishes has brought new and improved levels of performance to moisture management, in both waterproof, windproof, and breathable fabrics.

Gore-Tex® uses a patented membrane made from expanded polytetrafluoroethylene (PTFE) that adheres to a face fabric in 2-, 2½-, or 3-layer configurations. The pores in the membrane allow water vapor to pass through, but are too small for the much larger water molecules to pass back. Windproofing is achieved by air permeability, which shouldn't

be confused with breathability. Breathability refers to the ease with which water vapor or humidity pass through a fabric.

The company's newest technology, the Active Shell, is thinner and lighter than its predecessor. This new membrane is a porous shell that bonds directly to a fabric without an adhesive, reducing overall weight. It is estimated to be 38 percent more breathable than previous membranes, with over 9 billion pores per square inch.

Polartec's NeoShell® is another air-permeable membrane that, when viewed under a microscope, looks like a spider's web spun from polyurethane thread. NeoShell® has superior breathability (traditional membranes require a temperature differential to build up before moisture begins to move), and is available on knit and stretch fabrics.

A number of other companies offer waterproof fabrics too, including Pertex's Shield+, Evo's eVent, Patagonia's H2No®, Mountain Hardwear's Dry.Q Elite, and three different fabrics from The North Face—HyVent®, HyVent® Alpha, and HyVent® 2.5L Eco. All have a different approach to achieving waterproof, breathable, and windproof fabrics.

Weatherproofing finishes on soft shell fabrics and stretch outerwear fabrics work in a slightly different manner. Breathability is achieved through diffusion. When moisture and heat created by the body build up enough pressure, moisture vapor is forced through the fabric. Exercise is usually needed to generate enough body heat to initiate this reaction, making this type of finish best suited to endurance protective gear, as well as skiing, snowboarding, running, and other outdoor aerobic sports.

2.11
Wind tunnel testing determines aerodynamic values for performance garments. Here Sail Racing test their Fleet Windstopper® Jacket on one of the skippers of the SAP Extreme Sailing Team.

2.12
Polartec's Neoshell blocks out all moisture with its waterproof membrane; it is also breathable.

2.13
Dyneema® fabric, primarily used in sails, was used for these *Carbon Crystal Sails* to create an innovative installation by pioneer designer Greg Lynn. The exhibit, sponsored by Swarovski Crystals, was held at Design Miami in 2009.

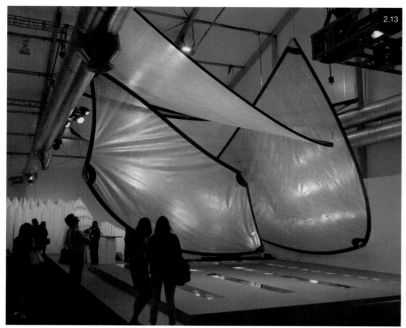

2: Safety and Protection

Abrasion resistance and slip resistance are also performance-enhancing features. Textiles with these features optimize user safety and productivity when used on protective apparel, and they improve performance when used for competitive sports apparel. Durability of a textile is key to the protection of the user. This includes prolonging the life of the textile in terms of strength, tearing, snagging, friction resistance, flame retardancy, and the ability to withstand assault from chemicals and radiation. Fibers and fabrics that are used to protect against extreme environmental conditions have been engineered with multiple properties to combat these harsh conditions. From body armor to fabrics that shield against unwanted radio frequency and electronic spying, protective textiles are becoming more commonplace in our everyday lives.

Shielding

In a world filled with wireless communication and data transfer, shielding fabrics protect us from invisible radio waves that surround us. The long-term effects of exposure to these waves is not fully known, and protection from them serves a dual purpose—preventative health benefits as well as protection from surveillance and tracking. Swiss Shield is a Swiss company that produces yarns used to create fabrics which can protect a wearer from electromagnetic fields, including cell phones and cell tower radiation, cordless phones, wi-fi, radar, microwave-oven leakage, and TV broadcasts, among others. Their fabrics are designed to be used in apparel, in industry, the military, and in home furnishings such as drapery and wall coverings.

2.14
Oracle Team USA won several national and international sailboat races using sails created from Dyneema®, a DuPont fabric that is lightweight, extremely strong, and highly abrasion and tear resistant.

2.15
The stiffness of the Dyneema® fabric is evident while the mainsail is being lowered. The 72-foot competitive race boat finished the race with a full 45-second lead.

Using extremely thin silver-coated copper wires spun with cotton or polyester yarns, their fabrics are woven into washable, comfortable, durable textiles that look and feel like normal fabric. Naturell™ is a translucent, unbleached ecological cotton fabric, designed for use in curtains, canopies, or clothing. Also for use in drapery and curtains is Daylite™, a highly transparent, open-mesh fabric with excellent air and light penetration. Its active components are 7.5 percent copper and 0.5 percent silver filament yarns, spun with polyester for strength. Another construction, Wear™, is a tight-weave cotton fabric with a gossamer-thin 0.02mm silver- and polyurethane-coated spun-in copper thread. More durable than Naturell™, this fabric can be used for bedding, clothing, and cell phone cases. These are just a few specific fabrics that are designed to shield users via apparel and home furnishings from unwanted exposure to radio frequencies.

Extreme environment protection

Fabrics made from fibers that form barriers against harsh chemicals, radiation, and heat are used in protective clothing for many industries. Ultra-high-molecular-weight polyethylene fiber is highly resistant to corrosive chemicals, odorless, tasteless, and nontoxic. These materials have extremely low moisture absorption and a very low coefficient of friction. They are also self-lubricating and resist abrasion. In fiber form, such material is sold under the brand names Dyneema® and Spectra®. It is spun through a spinneret and is used in body armor, cut-resistant gloves, climbing ropes and equipment, high-performance sails, yacht rigging, parachutes, and paragliders. Dyneema® is also used for puncture-resistant clothing in the sport of fencing, and other applications where abrasion resistance is critical.

Kevlar®, a synthetic aramid fiber blend created by DuPont, has been used in body armor for decades. It was first developed to replace steel in racing tires, and it now has widespread applications in clothing, accessories, and equipment because of its lightweight strength and ability to resist cuts and abrasion. Continuing innovations with the Kevlar® fiber mean that it is now found in everyday products ranging from iPhone cases to industrial applications, first responders' firefighting protection, and flexible military body armor.

Flame and high-heat protection

Many companies have developed fibers for use in fire protection and other high-heat applications. Twaron®, developed by the Dutch company AKZO Industrial Fibers, is a para-aramid and is used as a replacement for asbestos. It therefore has applications in many industries, appearing in such items as textiles and bulletproof body armor. The US chemical company Celanese has also produced many technically advanced fibers. Vectran™ is a manufactured fiber spun from liquid crystal polymer (LCP) and is often used in combination with polyester (as a coating around a

2.16
An eco-friendly cotton fabric, Naturell™ is used in curtains, canopies, and clothing, shielding against exposure to cell phone radiation, TV broadcasts, and radar.

2.17

The DuPont Thermo-Man® testing lab simulates worst-case scenarios that require the use of burn injury deterrents on Nomex®-insulated firefighter suits. The properties of these suits can also be extended to clothing for air force pilots and industrial workers.

2.18

The European Union funded the Fly-Bag project which incorporates auxetic textiles into the fabric of an airliner's baggage hold. Potentially this would prevent an explosive blast from spreading past the confines of the baggage hold.

Vectran core). One of Vectran™'s most notable uses was as a layer in NASA's spacesuit, the Extravehicular Mobility Unit designed by ILC Dover. It can withstand extreme heat (its melting point is 630°F (330°C), but it has progressive strength loss after 430°F (220°C), and is resistant to ultraviolet radiation. Finally, Nomex® by DuPont, is the most common material used in firefighters' kits in the USA today.

In 2013, the Swiss textile company Schoeller introduced pyroshell™—permanent flame protection for polyamide and polyester fabrics that can be applied to stretch fabrics. The combination of stretch with flame protection brings increased mobility and comfort to first responders and other workers in hot environments.

Schoeller has also developed a three-dimensional coating that creates an abrasion-proof, heat-resistant fabric called ceraspace™. The surface architecture of the fabric features raised bumps of hard duroplastic, a unique composition of special ceramic particles anchored in a polymer matrix. The ceramic particles are nearly as hard as diamonds and are firmly attached as a three-dimensional coating to the textile. This coating is highly resistant to elevated temperatures and works as a protective space barrier between the textile and a heat source.

Silica is another fiber used in fireproofing. When woven into yarns, it is similar to fiberglass and

was developed by Dow Corning as Beta cloth. The resulting fabric will not burn, and will melt only at temperatures exceeding 1200°F (650°C). Beta cloth was developed by a Manned Spacecraft Center team working on the Apollo/Skylab A7L spacesuit. It was implemented in NASA spacesuits after the deadly 1967 Apollo 1 launchpad fire, in which the astronauts' nylon suits burned through. Although it is an older technology, the use of silica is still viable today in the form of nano-coatings and impregnations.

Abrasion, cut, and penetration resistance

Many of these fibers not only protect against high heat but against cuts, abrasion, and penetration too, which is particularly important in military, law-enforcement, and industrial applications. SuperFabric® is a cut-, abrasion-, and penetration-resistant material manufactured by Higher Dimension Materials, Inc. As a technical fabric, it is created with a base fabric such as nylon, polyester, neoprene, or crepe, and then overlaid with tiny, hard guard plates in a specific pattern. The geometry, thickness, and size of the guard plates, as well as the base fabric, vary depending on the use. Customized and optional properties include flame resistance and specialized grip. It has a wide range of applications, including technical outdoor apparel, industrial safety and protection, motorcycle apparel

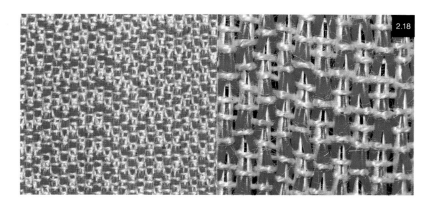

2.18

2.19

and accessories, heavy-duty diving suit protection, the military, and law enforcement.

Auxetix is a woven textile that grows thicker as it stretches, actively resists force, has the memory to return to its earlier state, and is flame retardant. Some of its many applications are hurricane protection, protective apparel for the military and law enforcement, concrete reinforcement for construction, and impact protection in extreme sports.

Aracon® is a metal-clad aramid fiber. Originally developed by DuPont as a Kevlar® product, it is now owned and developed by Micro-Coax. It combines the conductivity of an outer metal coating with the strength, light weight, and flexibility of an aramid fiber. The combined nickel, copper, and silver coating gives it a versatile set of attributes that have found applications in aerospace, commercial and military aircraft, communications, electronics, and conductive textiles.

Schoeller offers a line of protective fabrics that include keprotec® and reflex™. keprotec® is a base material originally designed for motorcycle racing and now used in motorcycle wear, workwear, active sports wear, gloves, and shoes. It is extremely abrasion proof, fall proof, tear resistant, and temperature resistant, and it offers a high level of comfort. It is a blend of Cordura® (high-tensile polyamide) and Kevlar®. Combined with different coatings and finishes, this fabric is designed to withstand the abuse of high-impact sports.

Reflectivity

Schoeller reflex™ fabric consists of a complex weave using a special reflective thread that provides extensive protection in poor visibility conditions. This yarn can also be combined with other fabric components for additional benefits, such as EN-471 yellow safety color, flame retardant, waterproofing, and active silver. When hit with direct light, millions of tiny glass beads embedded in the yarn create a brilliant reflection—a retroreflective reaction that makes it visible from up to 330ft (100m). It can be coated or bonded, and it can be used in both elastic and non-elastic fabrics, and in airy mesh constructions.

Scotchlite™ by 3M is another highly reflective material that has been adapted for use in textiles as both a transfer film or as a fabric, and works in a very similar way.

2.19
After the deadly 1967 Apollo 1 launchpad fire, NASA integrated the fireproof fiber silica into all of their spacesuits. Today, NASA has developed a polymer-reinforced Aerogel, stronger and more flexible than their current silica-based Aerogel insulation, which they hope to incorporate in all future spacesuits.

2.20
Biomimicry is the design of materials, structures, and systems based on nature. Dr. Anthony Brennan, working for the US Office of Naval Research, discovered that the skin of shark is a hostile surface for bacterial growth.

2.21
Using an actual impression of shark skin, he discovered that shark skin denticles are arranged in a distinct diamondlike pattern with tiny riblets that discourage microorganisms from settling. It simply takes the bacteria too much energy to grow on the non-smooth surface. Using this information, he created a pattern with a similar surface topography that has been proven to ward off bacteria without the use of antibiotics.

2.22
Sharklet Technologies has created the world's first technology to inhibit bacterial contamination through a physical pattern alone. Inspired by the texture of actual shark skin, the company developed a microscopic diamondlike ribbed pattern that mimics the natural pattern of shark skin and is proven to reduce bacterial growth.

3: Nanotechnology

Odor control and antimicrobials

Nanotechnology is the manipulation of materials at an atomic or molecular level. It is used in relation to textiles most often in the area of chemical finishes and coatings. Once fibers are spun into yarns and woven or knitted into piece goods, either the yarn or the fabric can be chemically treated with a finishing compound that embeds nano-particles in the finished material. The end results of these treatments range from superior moisture management and super-hydrophobic fabrics to fabrics that dispense medicines, aromatherapy, and anti-ageing moisturizers.

Silver has powerful antimicrobial properties; it creates an ionic shield that prohibits the growth of bacteria and fungi. X-Static® fibers are permanently bonded with a layer of pure silver, and are clinically proven to reduce bacterial growth on textile surfaces by 99.9

percent, including multidrug-resistant strains of bacteria such as MRSA and VRE. ActiveSilver™ from Schoeller is a finish designed to reduce body odor by inhibiting the reproduction of bacteria, mites, and fungi with a finish that permanently anchors silver salts to the fibers and yarns of the textile. Other companies offer antimicrobial and odor-control fabrics, too, including Nanotex Neutralizer, Agion Active™ by Sciessent, and Microban, among others.

Created for hunting apparel, ScentLok is designed to mask all human scent so that the hunter becomes undetectable to the animal he or she is tracking. It employs different technologies, each working to control odor but all doing it in a different way. First, each fiber is embedded with a carbon alloy that combines activated carbon, zeolite, and treated carbon. Activated carbon is proven to absorb 99 percent of human odor; the other alloys absorb

the remaining 1 percent. Bacteria, the cause of all odor, are attacked in three different ways: with the silver-thread barrier at the cellular level, an antimicrobial that kills bacteria as they hit the fabric surface, and by bacteria escaping along with water vapor as it evaporates from the body. Finally, the fabric has a textured capture technology that works to control the odor molecules with little branch-like arms on a polymer, similar to the seed of a dandelion. Once absorbed by the textured surface, the bacteria are held until the fabric is washed.

Soil and stain release

For the last five years the US Army Natick Soldier Research, Development & Engineering Center in Massachusetts has been perfecting a military uniform that doesn't need to be washed. This is now in the final testing stages. The "omniphobic" coating resists a wide array of water- and oil-based substances. To reduce odor, antimicrobial additives are incorporated into the finish that will be applied to the fabrics used in the uniforms. With over a million soldiers—including reservists and National Guards—on active duty in the US alone, each issued with five uniforms, this could save a lot of laundry.

Similarly, stain-resistant technology is being researched by Tong Lin and his colleagues at the Australian Future Fibres Research and Innovation Centre at Deakin University. They are working on a new multilayered silica nano-coating that will actively work to push off dirt, soil, and water. The coating works by layering positively and negatively charged layers of silica nano-particles that are stabilized with ultraviolet radiation, anchoring it to the surface of a cotton fabric. Their method waterproofs almost any organic substrate, so it can be used on wool, coconut, or hemp. Testing has shown that the coating has withstood assault from acids, bases, soaps, and solvents, and has survived 50 washing-machine cycles.

2.20

2.21

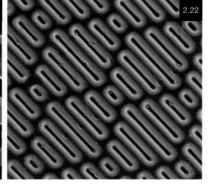

2.22

Another group of researchers at Monash University in Victoria, Australia, are working on coating natural fibers with nano-particles that will remove stains when they are exposed to sunlight. Nano-materials researcher Walid Daoud is experimenting with nano-particles of titanium dioxide radicals coated onto silk, wool, and hemp fibers. The titanium dioxide forms oxidizing radicals when exposed to UV and water. It also breaks down organic matter while leaving the natural fibers intact. This research could lead to future self-cleaning fabrics.

Shape memory: artificial muscles

An example of a nano-scale fiber is graphene, made from a single layer of pure carbon. This super material is 200 times stronger than steel, harder than diamond, incredibly light, highly flexible, and an excellent conductor of heat and electricity; it is only one atom thick. Its uses are just beginning to be explored. It has been suggested that graphene products will boost Internet speeds, increase microchip productivity (it is known to move electrons 200 times faster than silicon), and will be applied to touch-sensitive coatings and bonded to other fibers for improved strength and flexibility. Its extreme flexibility makes it most suitable for textiles. The carbon fibers that are used today are lightweight and very strong, with high abrasion resistance, but they are brittle. When bonded with graphene, however, they become incredibly flexible.

Materials scientists are also experimenting with graphene nanotubes. These consist of a single-rolled layer of graphene with a hollow core, one atom in diameter. These nanotube fibers form artificial muscle fibers and can crumple and relax in reaction to stimuli like heat or sweat. Scientists are currently using these fibers to create hybrid yarn muscles that can be manipulated to react in the same way that human muscles contract and relax to stimuli from our nervous system. Fabrics from these "shape memory" materials are self-powered intelligent textiles, and are starting to be used in both art and fashion today.

2.23
Utilizing the Nitinol shape-memory alloys, the scarf is able to respond to changes in temperature by either expanding or contracting, depending on the weather.

2.24
Graphene is a form of carbon that is one atom thick and is effectively two-dimensional. This computer illustration of the material shows its lattice-like molecular structure. The benefits of graphene include high conductivity, solar-charging capabilities and the fact that it is virtually indestructible.

2.25
The nano-scale fiber graphene is 100 times stronger than steel, but highly flexible and very light compared to other carbon fibers. There is speculation of its use in future cell phone models, notably that of Apple's iPhone.

2.26
The Oricalco fabric in this hood utilizes graphene nanotubes to create hybrid yarn "muscles". These synthetic "muscles" contract and relax when stimulated by electrical pulses. The reaction is similar to biological muscle that functions through stimuli received from the nervous system.

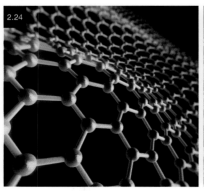

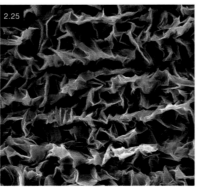

Nitinol, or nickel titanium, is another example of a material that changes shape. It belongs to a class of materials called shape-memory alloys. These materials have extraordinary properties, including shape memory, super elasticity, and a high damping capability. Such properties can be further manipulated and modified by changing composition, construction (mechanical workings), and heat. Nitinol shape-memory alloys undergo a phase change in their crystalline structure when they heat up and cool down. Its high-temperature state, called austenite, is stronger; its weaker state, called martensite, is super elastic.

When these alloys are in their martensite state, they are easily deformed and can be manipulated into a new shape or form. But as soon as the material is exposed to heat, its transformation threshold causes it to revert to austenite, and it recovers its previous shape and strength. Materials scientists have been able to isolate this transformation window and customize it to within a few degrees, or as wide as 100°C, depending on its end use. Nitinol is currently available in sheets, wire, ribbons, foils, tubing, and fibers that have been spun into blended yarns and has a range of uses from fabrics to medical devices.

Super hydrophobics

Super-hydrophobic fabrics go beyond simple waterproofing; they completely repel water and heavy oils. Any object coated with these finishes literally cannot be permanently in touch with liquid—liquids simply roll off the surface and leave no trace of their existence. A number of new super-hydrophobic coatings have been developed recently with the potential to make a large impact on industry and textiles. An example is NeverWet™ by the Ross Technology Corporation, a nano-coating that can be applied as a spray on hard and soft surfaces, or in the finishing process of a textile, leather, or other finished product. NeverWet™ not only prevents wetness, it also prevents corrosion, icing, and is self-cleaning. Applications for this product include clothing, shoes, sports, aviation, utilities, automotives, marine, construction, communications, electronics, and medical.

At MIT, a group of researchers have looked to nature for inspiration in creating a super-hydrophobic nano-finish that consists of tiny ridges on top of silicon that bounces water droplets off the surface of a fabric. Most super-hydrophobic finishes work by allowing only a very tiny part of the water to come in contact with the surface of a fabric or material. This finish has a different effect. In studying the patterns of certain butterfly wings and the surface of leaves, the researchers observed that these surfaces worked by reducing the water's contact time with the surface. When a water droplet hits the ridges it shatters axisymmetrically—into symmetrical shapes but in all different directions. Tests by the MIT team have shown that it has worked extremely well on common metals as well as the usual fabrics and polymers, and shows a 40 percent improvement on existing materials over other super-hydrophobic finishes. Applications for this product are roof tiles, glazing on tiles, and fabrics. The product could also be used as a coating on power lines to resist freezing and corrosion, and on jet engines and airplane wings to prevent freezing and ice buildup, among other industrial applications.

In ongoing research at China's Northeast Normal University,

 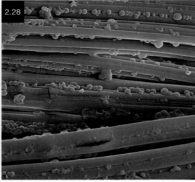

2.27
The effect of super-hydrophobic nano-coatings prevents wetness, corrosion, and icing.

2.28
The gold exfoliate, Nylgold, offers luxury dermawear with unparalleled skin protection and maintenance through the use of nano-particles. Nylstar, the developing company, plans to use the fiber in specialized activewear for high-end brands.

researchers have been working on a super-hydrophobic fabric that also has ultra UV protection, achieving an "ultimate" rating in UV protection status. They have coated cotton fabric with zinc oxide nano-rods as well as zinc oxide crystallites, giving the fabric a UV protection factor rating of over 100. The fabric achieves its super-hydrophobic property with a coating of silica on top of the zinc oxide, whose characteristics are enhanced by the combination of the two. At the University of Minnesota, another research team of chemists has created a nano-treatment that uses carbon nanotubes and Teflon to create a scald-proof coating for fabric. The mixture is so successful that it repels hot water, milk, and tea at 167°F (75°C).

Cosmeto-textiles

Using nanotechnology to embed cosmetic properties into textiles has the potential to become a multi-million-dollar business. With an ageing population, there is a growing demand for fabrics that have positive mental health and well-being attributes like providing moisture, aromatherapy, scent, and anti-ageing. A number of European textile companies have begun offering fabrics with these properties, and many such products have already been marketed, including yoga wear, shapewear, lingerie, and bedding.

Nylgold, first introduced in 2010, is a fiber with anti-ageing properties. It consists of a combination of gold and hyaluronic acid nano-particles that are adhered to polyamide 6.6 yarns during the spinning process. Hyaluronic acid is a mucopolysaccharide that occurs naturally in the skin. By helping the skin to retain moisture and manufacture collagen, it plays a key role in skin health and anti-ageing.

The French textile firm Euracli, which specializes in scented textiles, uses nanotechnology to create custom-scented creations for its customers. Working with the in-house perfumers, or selecting from their Aroma Library, customers can develop custom micro-capsule compositions that, when applied to base fabrics, are guaranteed not to change the color or texture.

Euracli also offers the EuraTex® line of cosmeto-textiles. These include a slimming treatment (SlimTex), a moisturizing treatment (HydraTex), a refreshing treatment (CoolingTex), and a firming treatment (LiftTex).

Cupron anti-ageing yarns use a patented process for incorporating copper oxide, which is known to be beneficial for both anti-ageing and odor control. Not only is it used in health and well-being but it also has applications in the military and medical fields. Cupron has been proven to promote healing in both skin health and wound closure. It has been applied to compression garments, diabetic socks, anti-fungal socks, and wound-care products such as gauze, dressings, bandages, and sutures.

4: E-textiles

Conductivity

As mentioned in Chapter 1, a large subset of smart textiles exists known as electronic textiles, or e-textiles. These fabrics conduct electricity, which means that some of them can store data, harvest energy, and generate and store power. They are used in a multitude of applications, including almost every area of wearable technology. Conductive materials such as stainless steel, carbon, and silicon are used in conjunction with glass, ceramic, and other fibers to create material systems within the textile constructions.

To create simple conductivity, metal fibers are twisted into yarns. These yarns are often sewn directly onto non-conductive fabrics to connect electronic components, in place of stiff wires. Fabrics woven or knitted from these yarns can gather information directly from electrical impulses from the user when worn directly on the skin. Information gathered from the environment can be transferred directly to embedded sensors, eliminating the need for wires. An example of this type of application is a heart-rate monitor embedded within a compression top or a sports bra.

2.29
This sweater's close contact with the wearer's skin allows conductive stainless steel yarns to gather information from the wearer's body. This data is then transferred to sensors embedded in the fabric.

2.29

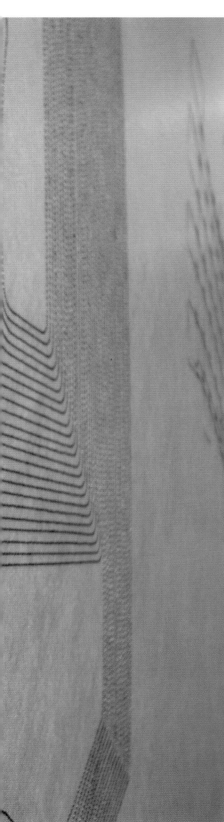

2.30
Conductive e-textile fabrics are used in most wearable technologies through storing data, harvesting energy, and generating and storing power.

2.31
The Forster Rohner embroidery machine creates custom e-textiles by twisting metal fibers into yarns that are embroidered and stitched on the textile surface.

2.32.
This medical-grade silver-plated fabric stretches in both directions and has a wide variety of applications in stretchy hats, socks, and gloves. It is excellent for creating electrode contacts and can be used in antibacterial wound dressing. The level of conductivity depends on which direction the material is stretched in. In one direction conductivity increases; in the other it decreases.

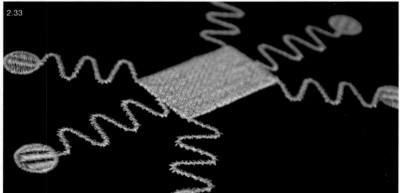

2.33
Fabric company Forster Rohner patented e-broidery technology that for the first time allows the integration of active lighting into fabrics without compromising the textile properties. Here sensors are embroidered onto the surface of the textile with conductive thread.

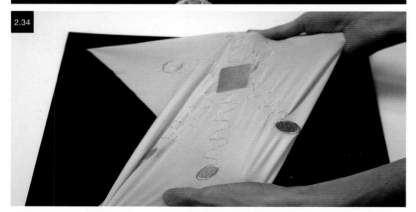

2.34
Combining electrical circuitry with the flexibility of an e-textile, these durable fabrics are both washable and conductive.

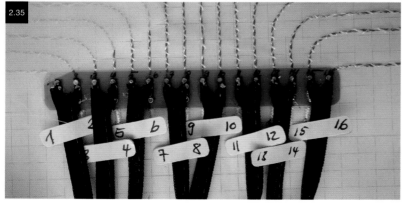

2.35
Not limited to fashion, the wearable sensors of e-textiles also offer caregivers the ability to monitor patients' behavior simply by way of specialized garments.

2.36

A 3D fiberglass woven fabric consisting of two deck layers bonded together by vertical piles in a sandwich structure. The material is light in weight, flexible, and ductile and is widely used in the composite industry.

2.37

This fabric has stainless steel on one side and polyester on the other, giving wonderful transparency effects. Double-sided fabrics are used for curtains, the decoration of shops, and in fashion.

2.38

The stretchable rubbery material created by researchers at the University of Tokyo can be used to place elastic circuits where previously only rigid circuits were possible.

By manipulating the fabric construction, different conductive effects can be achieved. A pressure-sensitive switching fabric is woven from two layers of conductive fabric sandwiching a layer of non-conductive material. When the fabric's two outer sides touch, the electrical circuit is completed. Double-faced fabrics (e.g., stainless steel on one side and polyester on the other) create translucent conductive fabrics with a soft hand feel for window treatments. Conductivity can also be achieved by coating yarns with conductive polymers. This is especially well suited to stretch fabrics, where conductivity is increased when the fabric is stretched in one direction and decreased when stretched in the other. Some of these fabrics made with metal yarns, have the added benefit of being sterile, and so can be used for wound dressings and other biomedical applications.

As the demand for more flexible and comfortable electronics grows, new conductive materials are being adapted and developed into textiles. Researchers at the University of Tokyo have developed a stretchable rubbery material that is highly conductive and can be used to build elastic circuits. The wires are made from a carbon nanotube-polymer composite that can stretch up to 1.7 times its original size without its performance being affected. The researchers foresee future applications in form-fitting athletic apparel that monitors

performance and body function, as well as in robotics. Layers of soft, stretchable circuits housing multiple sensors will replicate a "skin" for robots, giving them both a more lifelike appearance and sensitivity to touch.

Energy harvesting, storage, and generation

The idea of harvesting, storing, and generating energy from the environment and our bodies using textiles is fascinating, and a number of teams around the world are researching this field. One is Power Felt, a project at Wake Forest University in North Carolina, which will power small electronic devices by harnessing the energy released with fluctuating temperature changes in our bodies. The researchers have encased carbon nanotubes in a flexible plastic that is then entangled into a felt. The energy generated from shifting temperatures creates a charge strong enough to power a cell phone, small medical devices, or wearable sensors.

Energy storage is a key area of exploration. A number of teams in textile research are focusing their work on developing textile batteries. Wearable technology depends on a constant and reliable power supply. Today, most wearable electronics depend on a rechargeable lithium-ion battery. The development of power-supply technology hasn't kept pace with that of the components, leaving the user in the position of having to remove a device from their body in

2.36

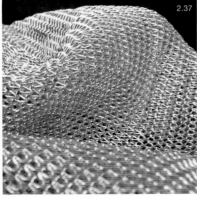
2.37

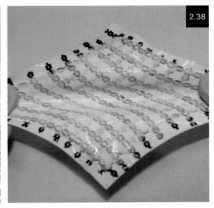
2.38

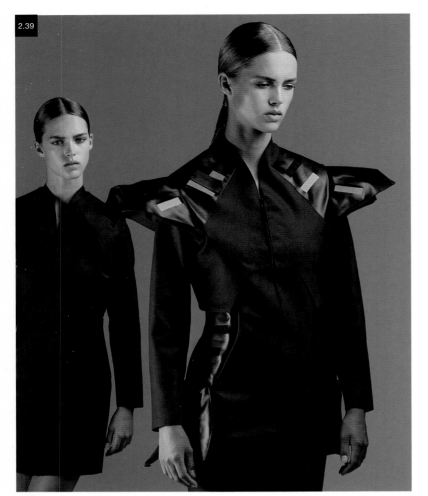

2.39

Everyday accessories, like this solar-powered outerwear by Pauline van Dongen, and other objects such as handbags and shoes, have the ability to harvest and use energy from the sun, the body's heat, or movement.

2.40

One of the benefits of flexible textile solar cells is the creative liberty they give designers. By not limiting its design to flat surfaces, designers have unlimited options in utilizing the material in their work while maintaining the conductive function of the e-textile.

2.41

Nanotechnologists at NASA have developed a textile that stores data similar to a memory stick, but in a flexible textile structure. Still in prototype stage, the design can store a gigabit of data for 115 days. This schematic shows how each data storage particle is nested in the intercies of the weave. The idea is theoretically to allow any type of interactive tool or piece of furniture to respond to its user and develop memory, like a computer.

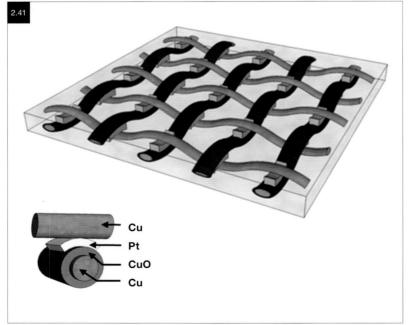

2.42

Forster Rohners Textile Innovation's e-broidery fabrics have many applications from textile lighting for safety or interior design, highlights for events and displays to stage costumes or sparkling evening dresses.

2.43

Open-work textiles, like this black lace, house conductive threads and LEDs and take a traditional lace fabric to a new dimension.

2.44

The LED components blend seamlessly into fabrics. Lights in this fabric are incorporated by adapting electronics into a traditional embroidered lace.

order to recharge it. In an effort to resolve this, a group of researchers at the Korean Advanced Institute of Science and Technology is working to develop a lithium-ion battery fabric. They have developed a fabric-based battery of nickel-coated polyester yarn as the current collector, polyurethane as a binder holding materials together, and a polyurethane separator. The resulting fabric is charged using solar cells and can withstand repeated folding and unfolding. Although it is still in the development stage it can scale easily and be worn as a wrist device.

A Japanese company has also been developing a solar energy-harvesting textile. Sphelar Power Corporation has created a prototype textile woven with 1–2mm microspherical solar cells. Unlike traditional flat solar cells, the microspherical solar cells capture rays from all directions, making them less dependent on the angle of the incoming light and more productive in terms of energy yield. Coupling these solar cells with the flexibility of a textile will allow solar harvesting in a variety of surfaces, including organic shapes, wearables, and those in movement.

Memory is another form of energy. Two NASA nanotechnologists from the Ames Research Center in California have developed a computer memory-storing e-textile. The fabric structure is a woven lattice of copper wires, where the top wire is left bare and the bottom wire is coated with copper oxide. At each intersection a small piece of platinum is inserted between the layers. The memory is stored in the copper dioxide by a process called resistive switching. This prototype is flexible, and can hold memory for 115 days 1/8sq. in. (1sq. cm) can hold over a gigabyte of information.

Illumination

The manipulation of color and light has fascinated artists and audiences alike for centuries. Fabrics that capture light and change color can capture our imagination, amaze, and mystify us. Photoluminescent inks, LEDs embedded in a woven textile, and fiber optics are all examples of how textiles can transform with light and color.

Lighting systems have begun to be incorporated directly into fabrics. Many artists and designers still create work by sewing LED components onto fabric with conductive thread, by weaving them into fabrics themselves, or by using conductive inks that connect them into circuits. LED components have become smaller and easier to use, with two holes that can be used to weave and sew them into projects by passing conductive thread directly through the component. But the newest technology is fabric that has integrated active lighting.

A number of companies are working on different ways to

2.42

2.43

2.44

2.45

Designer Zane Berzina developed *E-Static Shadows*, a woven electronic textile membrane that reacts to electrostatic energy. The final piece is created from hundreds of hand-soldered LED lights, transistors, and woven electronic circuits.

2.46

The textile circuit was woven on a large industrial Jacquard loom with conductive thread, then electrical components were hand-soldered into place. The soft light surface turns off when the membrane detects static charges.

2.47

The soft light installation registers the amount and intensity of the charges exposed to the sensory electronic textile membrane and translates them into a series of transient audiovisual patterns on the surface of the cloth. The project causes the viewer to think about the energy resources of our planet and investigates the human body as a generator of energy.

2.48

The expansive e-textile installation took over three years to complete and when electrostatic fields are detected creates the effect of "transient shadows". Viewers interact with the display which reacts both visually and sonically when the electrostatic charges of their skin or objects are encountered.

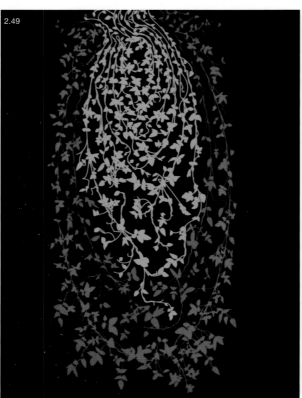

incorporate LED and OLED lights directly into textiles. Lighting effects can be achieved, for example, using embroidery, where LED components are applied to the surface of a body fabric as part of a decorative pattern, much like traditionally made embroidered laces. These new e-textile laces are made in a number of different ways: They can be embroidered, knitted, or made with tape. The LED components are applied within the decorative needlework with conductive thread. The Swiss company Forster Rohner has developed a patented e-broidery® technology, an LED lace fabric that is made in production fabric yardage.

Fiber optics are very fine, flexible glass or plastic materials that have the ability to transmit light. Fabrics made with fiber optic yarns have the ability to transmit a single LED light over their length, transforming a single spot of brilliant light to an illuminated path of light or an overall lit surface.

Photoluminescent fabrics increase their luminosity in response to their surroundings—a glow-in-the-dark effect. Inspired by photosynthesis, they absorb natural energy from the sun during the day, which is stored as electricity and then released as luminescence at night. These fabrics are extremely light stable and will not lose their illumination over many years of use. This makes them ideal for a range of safety and design applications.

Color changing

From the very beginning of the smart-textile movement, color-changing inks have been at the forefront of experimentation. In her landmark early work Maggie Orth (see page 134) experimented with electroluminescent inks in wall hangings. Today, artists and designers continue to explore light and color-changing inks in their work.

Thermochromatic inks change color with temperature. This can be

2.49
Created by UK design firm Loop.pH, the surface of this reactive window blind, *Digital Dawn,* is in constant transition as it responds to it's environment. It emulates the process of photosynthesis digitally with photoluminescent inks. A natural botanical pattern "grows," gaining luminosity as the environment slowly gets darker.

2.50
This piece explores how changing light levels in a space affect the inhabitants' physiological well-being. *Digital Dawn* uses the natural energy of the sun, storing electricity to illuminate the blind in the evening. The project explores the potential of smart fabrics to cross boundaries between physical and virtual spaces.

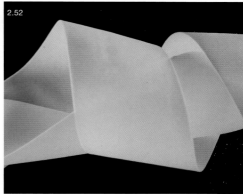

2.51

This pillow is a project designed by Linda Worbin while she was a doctoral candidate in Textiles and Interaction Design at the Swedish School of Textiles in Borås, Sweden. The use of e-textiles has lent itself to alternative assistance with patients suffering from dementia. Utilizing *Tactile Dialogues*, caregivers are able to interact with patients by way of vibrations in the material, enabling communication where conventional methods might not be possible.

2.52

Absorbing natural energy from the bright light of the sun, this photoluminescent ribbon produces a glow-in-the-dark effect by emitting a visible glow. With only a 5–30-minute exposure, the material will emit light for over 8 hours.

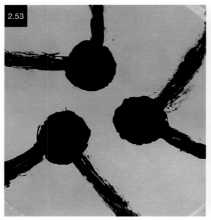

2.53

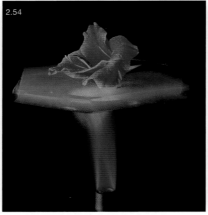

2.54

2.55

achieved with the introduction of a low-voltage electric current, so by varying the current you can also vary the color.

Researchers have also drawn inspiration from ocean creatures such as the zebrafish and the squid, which have the ability to change the color of their skin. A team at the University of Bristol in the UK has simulated these creatures' physical reactions with a mechanical process that they hope to apply to smart textiles using artificial muscles. Both of these animals have slightly different ways of changing color: The squid has pigmented cells that will expand and contract on command, rapidly growing very large or very small. The zebrafish, on the other hand, produces a liquid dye that squirts in and out of receptacles under its translucent skin, varying its color and pattern on demand.

By using artificial muscle fibers or dielectric—soft, stretchy, electrically activated—polymers, the researchers have simulated both of these systems. When a current is applied to the dielectric polymer, it expands and the appearance of the "color" grows. When the current is turned off, the color field decreases. The hope is that this technology can be applied to textiles that can adapt themselves to the look of their surroundings, leaving the camouflage obsolete.

5: Non-woven materials

Beyond traditional textiles, there are a number of non-woven smart materials being used in applications ranging from womenswear to body protection for contact sports and beyond. Many of these materials follow advanced developments in foams, films, and laminated composite materials. Experimentation by designers and fine artists have yielded cutting-edge projects that showcase the dynamic range of this group of materials. To create garments and other wearable equipment using these materials, specialized manufacturing techniques have been developed, including seam welding, laser cutting, and 3D printing.

Insulation

The Italian research lab Grado Zero Espace has been working on a collaborative project with the European Space Agency (ESA) called Safe & Cool. While working with ESA to advance the thermal and cooling qualities of textiles, the group developed aerogel, a fluid material made with nanogel. Originally created to insulate instruments in space, aerogel is virtually weightless and can withstand extreme temperatures. Grado Zero Espace went on to apply aerogel to other products, including the Quota Zero jacket and the Absolute Zero jacket, which could withstand temperatures down to -58°F (-50°C).

2.53

Biomimicry is the design of materials, structures, and systems that are modeled after nature. When looking to understand and recreate color-changing behavior, researchers from the University of Bristol, UK, are looking at creatures like the zebrafish and the squid, who can change their color from light to dark and back again. This image shows a shifting pattern in black pigment recreated in the lab to mimic these animals' ability to change color by contracting their muscles. The scientists used electric currents to simulate muscle contractions.

2.54 & 2.55

Aerogel is a synthetic ultra-lightweight thermal insulator. It is the lightest solid and the most effective insulating material in the world and sometimes referred to as "frozen smoke." It can insulate at up to -200°C and it only melts at 3000°C. In its purest form it can even float on air. Italy's Grado Zero Espace has used aerogel in garments with the development of the Aerogel Design System, a special thermo-isolating padding.

2.56

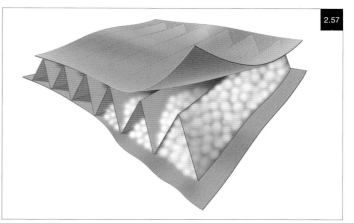

2.57

2.58

2.56
Thermoball™ insulation is made up of tiny spun polyester fiber balls.

2.57
Designers at The North Face designed a baffle system that suspends the Thermoball™ insulation in an even layer between the jacket's shell and lining.

2.58
Jackets constructed with Thermoball™ insulation are not only lightweight and incredibly warm but the insulation is superior to traditional down in that it continues to insulate even when wet.

In a further collaboration with both Hugo Boss and McLaren, they created special clothing for Formula One racing drivers and pit crew. Other applications for fully integrated cooling technologies were also developed for firefighters, first responders, steelworkers, oil field engineers, and motor racing athletes. Since then, aerogel has been used by many other companies and is considered the gold standard in lightweight extreme thermal insulation. It is often referred to as "blue smoke" or "frozen smoke," and is set to revolutionize many industries, from apparel to home insulation and electronics.

Clothing company The North Face has introduced Thermoball™ insulation. Unlike traditional synthetic insulation, which consists of long filament fibers laid down in flat sheets, Thermoball™ insulation consists of

very short polyester fibers that have been rolled into miniature fuzzy balls ranging from 1⁄8–1⁄4in (0.4–0.5cm) in diameter. These balls are then grouped into rows of interconnecting baffles, trapping air all around them. The advantage of the balls is threefold: They are extremely lightweight, they don't compress, making the garment very easy to pack, and because they are designed not to clump together, they maintain their insulating ability when wet.

Impact protection

Impact protection has been necessary since the early days of contact sports, but as the spectacle of sports has grown, so has the danger level. This has led to the need for better and better protection for the athlete, while reducing the feel and profile of any padding, and allowing for a greater

range of motion and mobility.

Traditional impact reduction consists of foam padding placed inside clothing, under hard plastic shells for helmets, and specialist protective equipment for football, hockey, lacrosse, and various motor sports. Foams are usually described in two categories: closed cell and open cell. Closed-cell foam has air bubbles completely enclosed within its structure; it is nonporous and is not breathable. Usually more dense, it is ideal for impact protection, seals,

and thermal insulation. Neoprene, ethylene vinyl acetate (EVA), and polyethylene are examples of closed-cell foams. Open-cell foam is created when the air bubbles burst during the manufacturing process, creating an interconnected cellular network. Open-cell foam is generally soft and compressible. It's breathable—water and air can pass throughout it with ease—and lightweight. It is often used for seating, packaging, filters, household sponges, and acoustic applications. Polyurethane foam and

melamine foam are examples of open-cell foams. Impact protection often uses a combination of foams—open cell for comfort, and closed cell to absorb impact.

6: Smart foams

The primary problem with any impact-protection gear is that it reduces an athlete's mobility. The thickness and placement of the hard plastic and foam of traditional equipment inhibited optimal performance. Athletes were often altering it, or opting not to use

2.59

Used in both commercial and noncommercial applications, D3O has received attention from many various outlets. For instance, the US Army's acquisition agency has funded the new D3O shock-absorbing helmet prototype which focuses on common blunt traumas to the head sustained by soldiers. Here the Demon Vest with D3O protects a mountain biker from impact without inhibiting a full range of motion.

2.60

This smart textile, Deflexion™, remains soft until struck by a high-impact force, when it instantly changes its makeup to protect the wearer from impact. Once the force is decreased the textile immediately returns to its flexible state.

2.61

There are two versions of the Deflexion™ technology, a 3D spacer technology and a thermoplastic technology. The 3D spacer textile is impregnated with silicone and is flexible, breathable, washable and durable. It performs well in a wide range of temperatures and works while wet. It is ideal for use in garments and is comfortable to be worn over long periods of time in both warm and cold conditions.

proper protective gear, resulting in injury and even death. However, a new group of smart foams has emerged to address this problem. These materials work to free the athlete from heavy, bulky, and awkward padding, provide a full range of motion, and thereby encourage the use of protective equipment.

Poron® is a microcellular technology that works to dissipate force from an impact over a wide surface area through phase change. Poron® XRD® is a soft, contouring open-cell foam that is breathable, and very effective upon high-speed impact. The material is soft while at rest and above the "glass transition temperature" (Tg) of its urethane molecules. When impacted suddenly or stressed at a high rate, the urethane molecules momentarily "freeze." Poron® has many applications, and in athletic impact protection has proven to be comfortable and lightweight. Because it is flexible, it frees the athlete from rigid, bulky, and constricting padding while still absorbing 90 percent of the energy when impacted at high strain rates.

Another smart, sensitive material used for impact protection is D3O. D3O is rate sensitive, which means that the stress versus strain characteristics are dependent on the rate of loading. D3O, also known by the acronym STF, is a dilatant or shear-thickening liquid, as opposed to a phase-change material. In a dilatant,

the viscosity of the liquid increases with stress. The process is controlled by the size, shape, and distribution of particles or colloids, while suspended in a liquid state but not dissolved in a solution. Shear-thickening behavior occurs when colloids change from a stable state to a state of coagulation, in which the particles lock together.

D3O was first developed by Richard Palmer and commercialized in 2006. An avid snowboarder, Palmer was inspired to find an application for D3O in sports apparel. "I recognized that if I could build a material which made use of those liquid properties but was no longer a liquid then that would be extremely useful in the context of snowboarding protection," he said. His solution was to transform the D3O goo into a foam-like structure which is flexible, retains its shape, and offers high shock absorbency. Beyond snowboarding, D3O has been used in a wide range of applications, including military helmets, sports equipment, workwear, footwear, medical, motorcycle apparel, and cases for electronics.

Working to create a comfortable material to reduce impact, Dow Corning developed Deflexion™, a smart textile that remains soft and flexible until it is struck by a high-impact force, at which point it instantly stiffens to protect against injury. Once the force is taken away, the material reverts back to its flexible state. The Deflexion™ S-range of

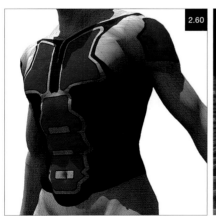

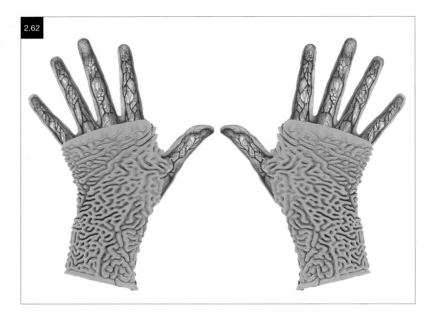

2.62

2.62
Israeli designer Neri Oxman has developed
a sophisticated 3D printing process that
allows her to vary the thickness and density
of materials as determined by the end use.
For example, Carpal Skin was developed
to protect against Carpal Tunnel Syndrome,
a painful condition of the hand and fingers
caused by compression of major nerves.
With Carpal Skin, a patient's pain profile is
mapped and then the skin is printed with
a distribution of hard and soft materials that
restrict movement in certain areas of the
wrist and fingers to bring relief.

2.63
Francis Bitonti used an experimental new
flexible filament and the MakerBot Replicator
2 3D printer to create *The Verlan Dress* for
his 2013 New Skins collection. Working with
a group of students from Pratt Institute in
Brooklyn, NY, Bitoni experimented with forms
in ZBrush and Rhino before realizing the
finished piece.

smart fabrics works with a 3D spacer
textile impregnated with silicone; it is
breathable, washable, durable, and
offers good protection. It performs
from -4 to +104°F (-20 to +40°C),
is flexible from -14 to +104°F (-10 to
+40°C), and works well when wet,
making it ideal for integration into
garments that are worn in warm
conditions or over long periods of
time.

Experimental research

There is some intriguing experimental
research being done at the
intersection of smart textiles and
wearable technology. There are
only a few projects but these are
important for their diversity and
direction. Computational graphics
and 3D rendering have begun to
intersect with actual form, creating
real objects through 3D printing.
Biology and textiles are intersecting
more directly in the lab, too, with
textiles and fashion textiles being
grown from biological materials. And
designers are fashioning textiles from
waste products and experimenting
with spray-on clothing. All of this work
cannot be discounted in a discussion
of smart textiles as it is so often
the conceptual exploration of the

unknown that leads to the next
breakthrough idea.

A number of multidisciplinary
designers have begun to work with 3D
printing to create incredible sculptural
forms for the body. Francis Bitonti,
a researcher at the Digital Arts and
Humanities Research Center at Pratt
Institute, created two iconic pieces
using experimental 3D printing—the
Dita Von Teese 3D dress (see page
30) and *The Verlan Dress*. Although
technically not fabrics at all, these
sculptural, computer-generated
wearable objects are so advanced
in their creative process and their
innovative approach to apparel that
they warrant a mention.

Leah Buechley, a pioneer in
wearable technology, has worked with
a group of students on the future of
textiles, focusing specifically on the
intersection of rich craft traditions
and new technology. Some of the
work that has been created under
her direction at the MIT Media Lab
has yielded some very interesting
concepts. For example, a student
has created a 3D-printed knit fabric
swatch. Using a multistep process, the
final result is a fully functional stretch-
knit swatch that is not created from
yarns.

2.63

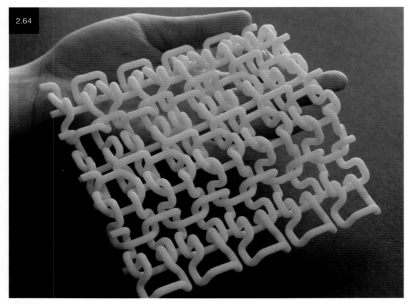

2.64

2.65

2.66

2.64
While studying at MIT under Leah Buechley and exploring new textiles, Alice Nasto created this 3D-printed knit swatch.

2.65
Through form experimentation, the New Zealand-based designer Earl Stewart, working in collaboration with Shapeways, created the XYZ shoe using a combination of 3D-printed nylon and traditional shoemaking materials.

2.66
Early versions of the shoe's form were realized through a 3D-printed architecture.

2.67
Biolace by designer and researcher Carole Collet are genetically engineered plants that produce textile and food, at the same time. "Biolace proposes to use synthetic biology as an engineering technology to reprogram plants into multi-purpose factories," explained Collet, who is a full-time academic and deputy director of the Textile Futures Research Centre at Central Saint Martins in London.

2.68
The project comprises four plants that are genetically engineered to deliver multiple functions. Here a strawberry plant with black lace growing from its roots yields black strawberries enriched with enhanced levels of vitamin C and antioxidants.

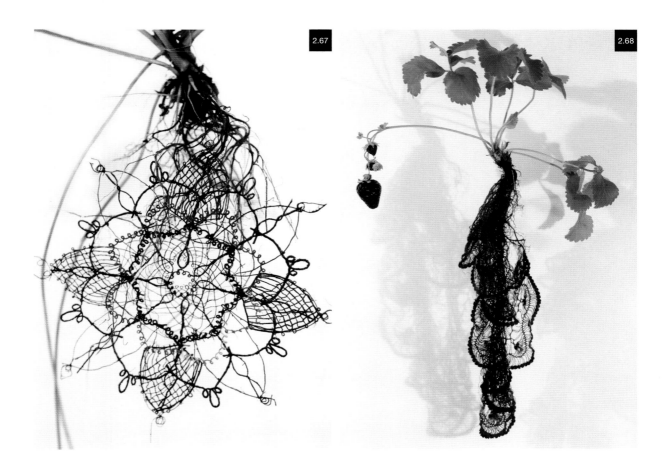

The effects of 3D printing are even starting to show up in footwear. The New Zealand-based company Shapeways, under the direction of creative leader Earl Stewart, has created the XYZ shoe, which uses a combination of 3D-printed nylon and traditional shoemaking materials such as leather and laces. These advances in 3D printing are just the first steps in what may become a revolution in materials and form experimentation.

Biolace, an experimental design research project by Carole Collet, is at the forefront of defining what textiles can and should be. Collet's work explores the boundary between biology and textile design, looking at the future of growing tomorrow's textiles. Meanwhile, designer Laura Anne Marsden looks at the other side of the biological chain. Her work focuses on "upcycling." Using refurbished consumer waste as her base material—namely plastic bags—she has developed a process called "Eternal Lace," which is carried out completely by hand.

For centuries, man has experimented to develop new and improved textiles—lighter, softer, stronger, smarter, waterproof, flameproof, bulletproof, more protective, conductive, color-changing, illuminating, and energy generating. If you can imagine it, it seems there is a way that it can be achieved. In the next chapter, we will look at projects by some of the world's leading creators, showing how they have applied these materials and pushed the boundaries of what we can expect from the fabrics that touch our lives.

PROJECTS

Designers, architects, and engineers are problem solvers. Where nothing existed, they create something new. They see a problem and solve it. They imagine something and build it. They observe a need and find a solution. To do this they have to be fearless in the pursuit of knowledge, always questioning, experimenting, and looking for a new material or a new way to use an existing one. The key to being able to generate creative solutions for design problems is to approach them with an open curiosity that comes from having a broad range of knowledge. You never know when and how a new idea will be generated. This is probably why you have this book in your hand: to learn about smart textiles. But there is so much to learn about such a highly technical field, how are you going to grasp it all and apply it your creative process?

This chapter looks at some of the ways smart textiles are being used today—whether it is in industry or in the studio. Starting with a survey of how smart textiles are currently being used will give you a foundation of knowledge to build from when you want to start a project that involves a new technology.

The following projects will serve as both inspiration and foundation and perhaps spark a solution to a nagging design problem you may be trying to solve. This chapter will explore a range of art, design, and scientific projects using smart materials and different manufacturing processes.

Light and Colour

To create unique and memorable theatrical experiences, performing artists use stage and costume design to enhance their shows, and by far the most popular smart textile design element used is the integration of light and color into fabric. There are a multitude of artists, designers, and researchers experimenting with LEDs, electroluminescent fabrics, and photonics to create sets and costumes. Many designs include programmable textiles integrated with LEDs that respond in time to music. Musicians and stage performers including the Black Eyed Peas, Cirque du Soleil, Lady Gaga, Katy Perry, and U2, among many others, have also used programmable electronic costumes that change color during their performances.

Two of the world's most watched televised events—the opening ceremony for the 2014 Sochi Winter Olympics, and the Super Bowl halftime show—have also used LED costumes that change color in line with choreography for amazing dramatic effect. The results created with integrated sound, light, costume, and set engage an audience, and create a connection between them and a performance.

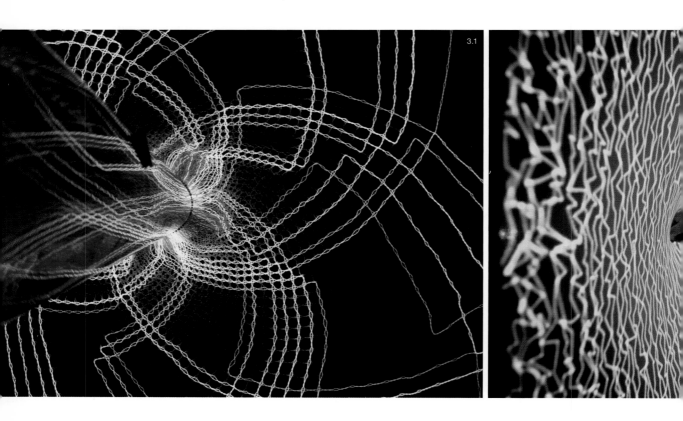

3.1

3.1

Sonumbra is a sonic umbrella of light that is animated from the sounds and movements of passersby. As a visitor wanders closer to the installation, they inadvertently become part of the composition.

3.2

Firewall is a combination of electronics, music, and a stretchable fabric interface. Interacting with the installation physically triggers audio and visual responses. The user can manipulate the kinds of sounds the piece will emit by varying the pressure into the membrane and adjusting the speed at which it is delivered.

3.2

Loop.pH, the London-based design studio, often uses smart textiles to create thought-provoking architectural installations in public spaces. Many of these installations use light to create visionary experiences that engage the viewer in an interactive experience. Loop.pH's work explores the social and environmental impact of emerging biological and technological futures, and the environments they create often synthesize living materials with digital tools. One piece, *Sonumbra*—commissioned by the Museum of Modern Art in New York—uses suspended electroluminescent fibers to form a huge umbrella-like shape that can offer shade during the day, and lighting at night. Lines of atmospheric light dance upon the shape in rhythmic patterns created by the movements of visitors; each individual's position is illustrated with a single strand of light, so that when a number of people move around the installation, a continuously changing lattice pattern is formed. *Sonumbra* responds to sound, transforming it into light and space.

The idea of using smart textiles to engage a viewer has also been used by many fine artists, especially those working with fibers, sculpture, performance, and installation. The supernatural ability of smart textiles to transform, change, and react has captured the imaginations of both artists and audiences alike. *Firewall*, by Aaron Sherwood and Michael P. Allison, is an interactive audiovisual installation that uses a stretched sheet of spandex (elastane) as an interface. The fabric surface can be manipulated to create music as well as firelike visuals. As you press into the membrane and move across it with more speed, the music responds with more intensity and expressiveness, becoming louder and faster, or softer and slower. The sound, coupled with the striking visual effects that appear across the surface of the fabric, create an immersive musical and visual experience.

Communication: Touch, Voice, and Data Transfer

Imagine being able to hug a loved one who is miles away. The sensation and intimacy of a hug transcends words and is a fundamental form of human communication. Today, wearable technology and e-textiles are making it possible to transfer touch as easily as voice, adding a new dimension to communication, and realism to long-distance relationships.

The strong and natural urge to connect with friends and loved ones drives so much of our lives; as a result, many of the world's greatest inventions have been around communication. Smart textiles are now giving us even more ways to explore and discover virtual physical connections, opening up new ways of looking at the world. When electronics are integrated into clothing, our garments become our interface with the Internet, dissolving the need to be constantly looking at, holding, and touching a screen.

Ping is one of these examples. This is a garment that uses your natural gestures to send messages wirelessly to your Facebook account. Created in 2010 by Jennifer Darmour—the design director of Artefact and author of electricfoxy, the wearable technology blog—Ping closes the communication gap by redefining the interface for social media. Lift the hood, work the zipper, drawcord, and button, or simply move by bending or twisting; all of these gestures trigger customizable messages to be posted to your Facebook page. As you go about your day, each of your regular movements can be constantly communicating your status, mood, or other information to your friends.

3.3

Ping is a garment that seeks to connect social life to social media life. This innovative design by Jennifer Darmour uses normal body gestures and garment functions like buttons and zippers to trigger posts and other commands on the user's Facebook page. The project explores using garments as an interface device.

3.3

When you get a ping back from a friend with a message or a comment, sensors built into the shoulder of the garment trigger a soft vibration. To be able to identify who is getting in touch, the Facebook app can be used to customize unique tapping rhythms for each of your friends or friend groups. You can, in effect, build a personal language that you control within your hoody to post and receive posts to your Facebook account. Ping stretches our imagination and illustrates where garments and accessories can evolve to house 3D gesture-recognition technology and environmental sensing. Soon our clothing will not only look and feel good, but it will also keep us connected without us having to hold a piece of technology in our hand all the time.

On a more intimate level, Australian fashion designer Billie Whitehouse has teamed up with Durex to create unique underwear that brings more than a simple conversation to your partner when you're apart. Billed as the future of foreplay, Fundawear is controlled by an app on your partner's cell phone. When activated at their end your panties pack a bit of a surprise, redefining intimacy for long-distance romances. This electronically enhanced underwear for both men and women uses conductive fabrics and laces in combination with lots of tiny, strategically placed vibrating sensors, all controlled by wireless Bluetooth interactive software. Remotely controlled sensors have thus created a growing market for erotic wearable technology.

The SMS Pillow

The international team at Philips Research, known for their advanced design concepts, have created some of the most innovative concepts in wearable technology. As far back as 2005, they first toyed with the SMS Pillow. Made from a photonic textile fabric, it can receive text messages and display them across its surface through an array of embedded LED lights. This pillow was the precursor to a number of projects that Philips and others have worked on in exploring the boundaries of alternative forms of communication.

Since that early project, Philips has created a padded flexible fabric called Lumalive. Woven with a matrix of full-color LEDs connected through a series of conductive fabric patches and threads, it is robust enough to withstand washing. Similar in concept to the SMS Pillow, but more

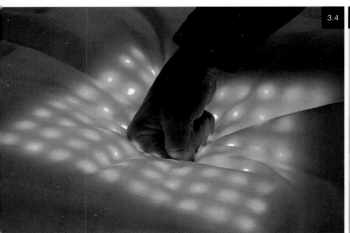

advanced, Lumalive was made with the interior lighting and advertising industries in mind.

Interestingly, three investigators, Sylvia Cheng, Kibum Kim, and Roel Vertegaal from the human-computer-interaction lab at Queen's University in Canada, have concocted a fun twist on the classic game of tag using shirts made with touch-sensitive Lumalive fabric and proximity sensors. The game, called TagURIt, is played with three people, a chaser and two target players. In the game, virtual tokens that light up on the chest of each of the players' shirts can jump from one shirt to another with when the chaser gets too close. Like a game of keep away, combined with tag, the chaser tries to catch the player with the token to gain points. When the chaser gets too close to the target player, the token can jump to the other target player's shirt. The longer a player holds the token, the more points they gain. TagURit is the first of many interactive games that will use e-textiles in interactive gaming and communications.

The "Internet of things"

The "Internet of things," a phase coined by British technological innovator Kevin Ashton in 1999, describes a world where physical objects are seamlessly integrated into the information network, and where businesses can interact with these "smart objects" via the Internet. These smart objects will have the ability to automatically transfer data over a network without requiring human-to-human or human-to-computer interaction. This is an area that is experiencing particularly rapid expansion in the field of interior furnishings. Smart appliances, smart windows

3.4
Philips Research has created an intelligent and interactive pillow. It can receive and project text messages across its face using a set of LED lights in the photonic textile fabric.

3.5
Philips Research's Lumalive fabric wall is a smart and colorful wall that can create expressively lit environments, it can be programmed to display everything from an evolving light show that shifts over time to advertisements.

3.6
Annette Douglas with researchers from research institute EMPA has developed new sheer fabric that lets light in while keeping sound out.

3.7
TagURIt is an electronic game of tag that uses proximity sensing and touch-sensitive Lumalive displays on garments. The game demonstrates how interactive e-textile displays can allow human contact even at a distance, turning the virtual into the physical.

and doors, smart lighting, smart HVAC (heating, ventilation, and air conditioning)—smart almost everything—will soon be able to create wirelessly integrated home and office environments that are more convenient, efficient, sustainable, and even safer.

A combination of smart textiles and embedded sensors will monitor activity in our surroundings to maintain constant and efficient temperature, adjust lighting as needed, conserve energy usage, and control sound. Artificial muscles made from smart polymers will be used to open and close windows as temperatures change. With little to no electricity needed to operate these smart systems, energy will be conserved in the areas of heating, cooling, and lighting.

Noise is annoying and disruptive. It breaks your concentration, reduces productivity, interrupts communication, and zaps your energy. If you have ever been in a busy place where a lot of people are speaking at once, you know how difficult it is to communicate, even with someone right next to you. Sound-absorbing materials therefore play a key role in creating successful environments, whether it's an office space, a restaurant, or even an urban residence. Interior architects and designers have long

considered the need to control sound in creating both commercial and residential interiors to improve quality of life.

Traditionally, sound-absorbing window coverings are opaque and padded with materials that will physically absorb sound—think heavy, theatrical velvet drapery. As sound waves pass through the heavy fabric, friction dampens the waves and the fabric absorbs the sound. However, window coverings like this also block out most natural light. In response to this problem, renowned Swiss textile designer Annette Douglas has developed a collection of groundbreaking new fabrics called "Silent Space." These fabrics look like traditional sheer drape panels that allow natural light to stream in, but they have the ability to absorb five times more sound than traditional sheer-curtain panels.

The "Silent Space" collection is the result of a research collaboration between Douglas and the Swiss Federal Laboratories for Materials Science and Technology (EMPA). The team used computer models to analyze the sound-absorbing qualities of different yarns and fabric constructions, and eventually chose to develop a fabric with multiple modified polyester yarns. The result is a lightweight, translucent, flame-retardant fabric specifically designed to absorb sound—perfect for acoustic curtains. Since its introduction in 2011 the collection has received many prestigious design awards, including the Red Dot "Best of the Best" award in 2012, and the Interior Innovation Award in 2013.

Smart carpets

Innovations in interior product design are also advancing in flooring. For example, smart carpets embedded with an optical fiber will soon be able to detect your gait and determine if you have fallen. Researchers at the University of Manchester in the UK have been working on carpeting that is woven with smart optical fibers under its pile layer, creating a 2D plane that acts as a pressure map. The gait of someone walking on the carpeting can then be identified from the data collected by this map, and analyzed to detect anomalies such as a limp or a deterioration in mobility—useful information for doctors diagnosing chronic issues related to ageing and diabetes.

The carpeting can also determine if someone has stumbled and fallen, and can be programmed to send out an emergency signal if the person doesn't immediately get back

3.8

3.8
Philips is collaborating with carpet manufacturer Desso to create a smart carpet embedded with LED lights that projects important messages and signs right at your feet.

3.9
Smart carpets can be used to create ambient lighting in commercial interiors as well as to control traffic through public transportation facilities, such as airports and train depots.

3.10
The use of LED lights in the floor is thought to more readily catch people's attention than signs hanging from the ceiling. The LED lights are woven into the fabric and create illuminated messages at ground level.

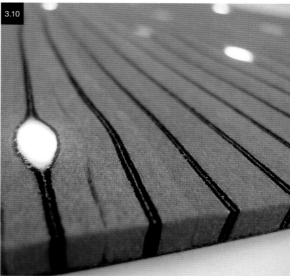

up. Estimates are that 30–40 percent of older people who remain living in their homes fall each year; this is the most serious and frequent accident in the home, and accounts for 50 percent of hospital admissions for people over 65. The carpeting can be retrofitted into someone's home at low cost, so could be particularly useful for an ageing population and for those with long-term disabilities. It even has the added benefit of being able to detect intruders and sounding an alarm, and could also be used by physiotherapists to map changes and improvements in their patients' treatment.

Smart carpeting is also being developed to replace overhead signage in public places. Philips has worked with the major carpet manufacturer Desso to develop high-quality commercial carpeting that has LED lighting woven into its construction. This light-emitting carpeting can be programmed to display important messages and symbols in multiple languages. The target application for this new venture is high-traffic public spaces such as airports and offices. Passing through a busy airport terminal, a light-up sign in the carpeting guides you on your way to connecting flights, baggage claim, or exits. When unlit, the carpeting either takes on a normal appearance, or the LED light glows softly. Light-up signage in the carpeting catches people's attention more readily, since most people look down when they are walking.

Shape Shifting

The desire to express abstract ideas visually has long driven the creative process, and now, with the introduction of smart textiles and wearable electronics, artists and designers are embracing these materials in new works that explore the intersection between our bodies and technology. Whether it's to explore new ways to visualize our environment or simply to capture our attention and entertain us, their work is the wedding together of art and science.

How can a fabric move and change shape? It may sound like science fiction but it's a reality. Artificial muscle fibers contract and expand in the same way as a human muscle. Triggered by heat, these small needlelike materials are stable until they reach a pre-programmed reactive temperature, at which point they transform. They will continue reacting until

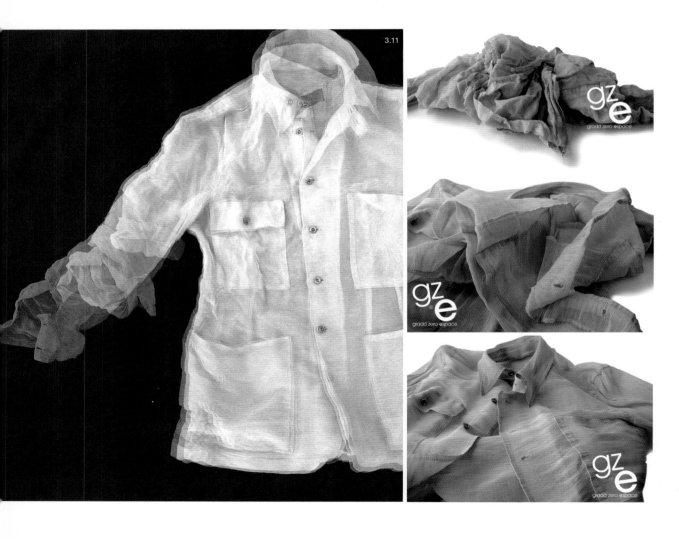

This shirt made from the shape-memory fabric Nitinol transforms from a relaxed to a compressed state when triggered by a change of temperature. The fabric can be programmed to an accuracy of two degrees and has been used for everything from medical and health products to performance activewear for severe weather conditions.

3.12
Grado Zero Espace's K-Cap face mask and headgear provides the ultimate protection from weather obstacles for extreme mountaineers and travelers alike. Its shape-memory membrane and bi-elastic fabric help it achieve a completely flexible but also memory-dependent fit.

an equilibrium temperature is achieved again. These fibers are twisted into yarns that are then woven into an artificial muscle fabric, such as Grado Zero Espace's (GZE) Oricalco.

GZE is a world-renowned Italian industrial design and engineering firm. They developed Oricalco through the Technology Transfer Programme Office (part of the European Space Agency), which encourages and facilitates the use of space technology for non-space applications. The fabric uses Nitinol, a shape-memory alloy that, when exposed to heat, has the ability to recover any shape with which it has been pre-programmed. To demonstrate the fabric's unique ability to transform itself, GZE created the Shape Memory Shirt.

The shirt, which appears to be a traditional collared shirt with patch pockets, has pre-programmed sleeves that automatically shorten as the room temperature heats up. Once the room returns to a normal, comfortable temperature, the sleeves revert to their previous shape. The beauty of the Nitinol fibers—the artificial muscles—is that they can be programmed to trigger at almost any temperature, and to form any shape, and can repeat the process many times. Shape-memory alloys are very exciting new materials, and Oricalco is the first time that Nitinol has been used in a textile.

Shape-memory fabric

The K-Cap, also designed by GZE, uses shape-memory fabric to create a uniquely warm balaclava for high-altitude mountaineering. The cap is formed from two layers of fabric that work in tandem. The inner one is a shape-memory fabric that shapes to the form of the wearer's head as the temperature drops, creating a seal that mimics human skin. The second layer has two pieces of bi-elastic fabric, a knitted fabric with two layers of stretch yarns working together to create equal forces in both longitudinal and transverse directions, allowing for complete freedom of movement. Unlike traditional hoods, which remain stationary when you turn you head, the K-Cap, when activated, becomes part of you, turning with your head as you shift your gaze. This offers great improvements in visibility over standard woven hood constructions.

GZE has also applied its design acumen to the S1 Sailing Suit for extreme ocean sailing. The suit's fabric uses a shape-memory membrane to change its permeability and adjust for breathability according to thermal conditions. A shape-memory alloy in the fabric's construction triggers a change in the weave. When it's cooler, the fabric's composition becomes tighter; when warmer, it becomes more open. Not only does the open weave improve breathability and maintain a constant body temperature, it also reduces the buildup of perspiration, which is very destructive to both fabric and finishes when it decomposes.

   The suit has other features to assist a wearer in a marine environment, too. Shock-absorbing pads made from a visco-elastic polyurethane memory foam are built in to protect those areas of the body most likely to face high impact, while the glove and sleeve are combined into a convertible one-piece unit to prevent the loss of a glove in adverse conditions. All the suit's seams are sealed with a patented Liquid Shell treatment that preserves flexibility while providing optimal water resistance in demanding sailing conditions. And finally, for safety and increased visibility, the back of the jacket is outfitted with an electroluminescent film, which can be operated from inside the pocket, lighting up the back of the suit in severe weather or darkness. New materials and

3.13

3.14

gz.e

grado zero espace

S1 suit

innovative design features work in concert to create one of the safest and most comfortable suits for extreme sailing.

Slow Furl

Traditionally. architecture is static. But life is full of movement. *Slow Furl* was a research project carried out by Mette Ramsgard Thomsen, in collaboration with Karin Bech of the Center for Interactive Technology and Architecture in Denmark, culminating in an architectural installation at the Lighthouse gallery in Brighton (UK) in 2008. The project explored the realm of movement within the confines of architecture, questioning how we perceive our built environments, and consisted of a textile membrane skin covering a dynamic armature. The skin was sewn tightly in some areas and draped more loosely in others, allowing for movement created by robotics, which were connected to the underlying kinetic structure. The textile membrane was embroidered with conductive threads that served two purposes: They connected soft switches that controlled the continual movement of the armature, and they added surface interest to the membrane.

3.15

3.13
Grado Zero Espace's S1 Sailing Suit is engineered to be a multi-purpose marine suit and is fully equipped for protection from the harsh conditions of cold-weather sailing.

3.14
Adjustable straps at the wrists allow the glove to be tightly fastened onto the hand, or when removed, to be tightly fastened to the arm.

3.15
Extra padding at specific locations, such as the back of the arm and shoulders, protects from high impact.

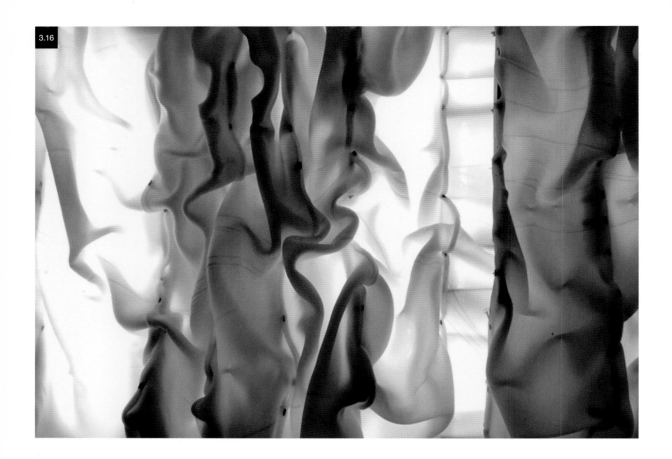

3.16

One of the defining features of *Slow Furl* was its slow—almost imperceptible—pace of movement. The team also discovered that the movement of the structure was much more noticeable without the skin. During the design process, the armature was programmed to move slowly but perceptibly, but when the skin was applied the movement become almost impossible to see, challenging viewers to question their ability to sense movement. The final installation left peepholes in the skin, so that observers could see the movement of the armature and compare it to the movement of the skinned surface.

This experimental installation was only the beginning of a conversation. Development is now underway for smart building materials able to actively monitor and control interior environments without electricity. These will help manage power consumption by maximizing solar and geothermal heat using a range of possible techniques, such as bending slowly toward the sun to maximize energy harvest.

3.16
Slow Furl is a room-size textile installation that reacts to movements and rhythms. Robotics with a textile skin create a slow-moving, transformative wall.

Scent and Sound

Aromatherapy—using scent to improve health and general well-being—is a medically proven therapy used by millions of people. Designer and researcher Dr. Jenny Tillotson has first-hand experience of its positive effects and has spent a remarkable career developing what she calls "an extra second skin" of scent on clothing. This scent layer is intended to add healing benefits to your wardrobe. Her "FreeBee" project, touted as changing the life of "scentient beings," is about "living fabrics" that deliver scent. Much of Tillotson's work as a senior research fellow at Central Saint Martins in London, in conjunction with researchers at Cambridge University, focuses on fusing fashion with technology and aromachology, the science of scent.Tillotson explained in her 2011 TEDxGranta talk that although scent is one of the least researched of the five senses, it has a direct pathway to the brain. It can be very evocative and has a direct pipeline to our feelings, likes, and dislikes. It also has distinct mood-enhancing effects.

3.17
Dr. Jenny Tillotson's *extra second skin* dresses are created from fabrics that release specific aromas in order to generate emotional reactions.

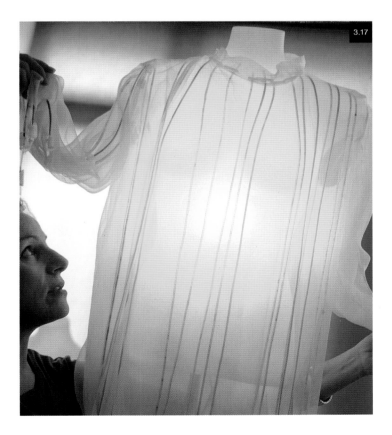

3.18

3.18
The Climate Dress senses pollutants in the air we breath and visually displays the data in real time with an ever-evolving display of lights. This detail of the embroidery shows the LED lights which are powered by soft circuits and conductive threads.

Tillotson creates garments from fabrics that have embedded scents designed to reduce your heart rate, have a positive effect on performance-induced stress, and work toward balancing the nervous system. Mental health issues are a growing concern across the globe. It is estimated that as many as one in four people in the UK, for example, struggle with mental health issues, including depression and anxiety, and nearly half of the world's population suffers from sleeping disorders. Using a combination of smart fabrics and wearable technology, Tillotson works to create garments that will deliver a personalized scent experience to the wearer. Her hope is to augment traditional treatments for mental health-related issues with her jewelry and clothing.

The Climate Dress, made by Danish design company Diffus in 2009, in collaboration with Forster Rohner Textile Innovations, reacts to pollution in the air by sensing high levels of carbon dioxide. What appears to be a beautiful jeweled evening dress is revealed upon closer inspection to feature embroidered beading made up of conductive threads that light up in response to CO_2 concentration to produce a range of different pulsating light patterns.

Movement and Temperature Regulation

Today, amateur and professional athletes alike are using smart textiles to gain a competitive advantage. These fabrics can help in the quest to become faster and stronger, and to excel in adverse conditions such as extreme temperatures, rain, and snow. Some of these technologies are even trickling down to our everyday clothing. Waterproofing, heat retention, and moisture management are now incorporated in underwear, outerwear, and everything in between. The use of smart textiles is particularly pervasive in active sportswear, and the sports-minded customer, eager to purchase clothing with these features, is not averse to their price tags. But as more and more companies start using smart textiles and adopting new technology, and the fabrics become easier to manufacture, prices will come down, allowing these fabrics to be adopted across the garment industry.

Controlling body temperature is key to peak performance for sustained periods of time, and many companies have moisture-wicking technologies that work with the body's own highly efficient evaporative cooling system. They work by wicking moisture to the surface of the fabric to provide a cooling and drying effect. Recently, though, a breakthrough smart fabric has changed the game by reacting with perspiration rather than wicking it away.

Columbia Sportswear's Omni-Freeze™ ZERO fabric is proven to lower your body temperature through a molecular change in its construction. The polyester wicking base is

3.19
The hydrophilic polymer rings become evident in Columbia's Omni-Freeze™ ZERO garment as the wearer begins to perspire.

3.19

embedded with thousands of $\frac{1}{8}$ in (3mm) super-absorbent blue polymer rings (more than 41,000 on a men's medium-size shirt) that swell into donut shapes when they encounter perspiration. This transformation process requires energy, which the fabric obtains from the athlete's body heat. Testing has proven that Omni-Freeze™ ZERO shirts are up to 10°F (5.5°C) cooler against an athlete's skin than those created from any other wicking fabric, making these the first shirts that actually lower your body temperature as you exercise.

Staying warm

Just as important as keeping cool is staying warm. Silver-plated polyamide threads are now being knitted into underwear and base-layer garments that warm up the skin directly. These are powered by a battery-controlled unit that is about the same size and shape as a cell phone and lasts for up to six hours at a time. Because the silver-plated polyamide is an excellent conductor, no wires are needed to spread the heat through the garment, and the fabric is durable enough to withstand machine washing.

Jumping to space age technology, the Italian company Corpo Nove, working with the European Space Agency, created the Absolute Zero jacket in 2003. What makes this jacket special is the insulating material it uses—aerogel, the material developed by NASA and used to insulate space

3.20
WarmX® is self-warming underwear. A silver-plated polyamide is knitted into the WarmX® Underwear, making direct contact with the skin and pumping energy through the garment to heat the wearer's body.

3.21
The Absolute Zero jacket insulates the wearer using Aerogel, an incredibly lightweight insulator first developed by NASA.

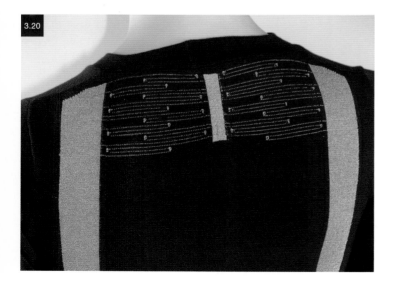

3.20

probes sent to Mars. Aerogel is a silicon-based substance with a spongelike, open-cell foam construction that is 99.8 percent empty space, making it the lightest solid material in existence. All that space enables it to trap and warm air more efficiently than any other known material, creating an incredibly lightweight and energy-efficient form of insulation. The Absolute Zero jacket has led the way to other innovations in lightweight insulations.

Extreme sports and X-Games enthusiasts, or even just typical teenagers who like the outdoors, are extremely hard on their clothing and need it to stand up to the repeated falls and scrapes encountered in climbing, hiking, skiing, snowboarding, and other winter activities. The Mithril jacket might be just the answer. Made with Kevlar®, the synthetic bulletproof polyamide fiber used in body armor, it is wind resistant, water repellent, and built to stand up to the nasty spills and everyday abrasion of extreme sports. The jacket is constructed with sealed seams to keep water out, and the soft-shell fabric is flexible and comfortable. It is designed to last a lifetime.

Off the slopes, smart fabrics are showing up in everyday clothing, and niche companies have begun to create specialized clothing that addresses the needs of customer groups as diverse as bike commuters, pedestrian-safety advocates, and urban nomads.

Phase-changing material

Ministry of Supply, the Cambridge, MA, company, was started by four MIT graduates who wanted a dress shirt that would stay fresh and dry all day long, despite temperature fluctuations in the office. Their Apollo dress shirt is made from a proprietary knitted yarn containing a phase-change material. It regulates your body temperature by storing body heat until you need it again to warm up. Phase-change materials store, release, or absorb heat as they oscillate between solid and liquid form, giving off heat as they change to a solid state and absorbing it as they return to a liquid state. Think of a sweltering July day where you go from freezing office building to scorching sidewalk to the staggering heat of the subway and back. Or even just the fluctuations you experience in body temperature when going from your desk to a client presentation. Since its initial

3.22
The Mithril Kevlar® jacket is extreme sportswear that remains lightweight, flexible, and durable while providing wind, water, and impact protection.

3.23
The H2 Flow jacket, by Helly Hansen, is a lightweight jacket for cold weather climates capable of maintaining consistent body temperatures using a series of specially designed internal pockets. Closing the pockets captures heat and warms the wearer; opening them does the opposite.

3.24
The 3-Season Suprema weatherproof pant by Makers and Riders doubles as a casual garment but is also action-ready outdoor bike gear. It is the ultimate in commuting apparel for the urban bicyclist.

Kickstarter launch of the Apollo shirt line, Ministry of Supply has introduced temperature-controlling undershirts, socks, and pants.

Urban outdoor brands Makers and Riders, Mission Workshop, Outlier, and Aether all market clothing that is equally suited to the boardroom and the bicycle commute, creating casual-styled designs that hide the sort of high-performance features usually found in competitive athletic clothing.

For example, Makers and Riders' best-selling bike pant, the 3-Season Suprema weatherproof pant, is breathable, waterproof, and flexible, making use of a new breed of soft-shell fabrics that are 100 times more breathable than other waterproof textiles on the market. The pant has five-pocket jeans styling, with trouser-style welts in the rear, and hidden internal hand pockets. Makers and Riders also worked to develop a unique gusset pattern in the crotch to enable complete range of motion while providing a slim-fit silhouette.

The Sporty Supaheroe Jacket by the Austrian company Utope uses a combination of wearable technology and intelligent materials like wind- and water-resistant breathable organic cotton and flexible LED circuits. This is not a typical illuminated cycling jacket. It has the ability to communicate with your mobile device to relay information on location, gesture, and speed. Its main lighting system, which consists of a series of flashing LED lights, is controlled by integrated sensors that track the cyclist's movements and body

3.24

position, offering improved visibility and after-dark safety. The entire wearable electronic system is powered with a rechargeable battery, and is built on a stretchable circuit panel housed between lining and outer layers of smart fabrics that are completely breathable, water resistant, flexible, and washable.

The jacket itself combines intelligent fabrics known for their cutting-edge technical performance. The main outer layer is EtaProof organic cotton, an incredibly strong and highly durable fabric, created to withstand military demands. Inset into the outer layer are translucent sections of fabric that allow the LED components to shine through when the wearable electronic system is turned on. These fabric inserts are stretchable, waterproof, and wind resistant. Finally, the lining is a viscose fabric combined with a soft-shell stretch fabric across the back and in the underarms to enhance moisture control and breathability while allowing a full range of motion. The Sporty Supaheroe jacket was awarded a Red Dot "Best of the Best" award in 2013.

For an athlete, measuring, monitoring, and comparing the results of physical performance are all part of training. Today, wearable technology allows a coach and trainers to see an athlete's performance statistics, and this information can even be shared with fans. In 2011, Under Armour

3.25 & 3.26
The Sporty Supaheroe jacket by Utope uses an array of LED lights to make nighttime bicycling and other activities more safe. It also has an appealing aesthetic.

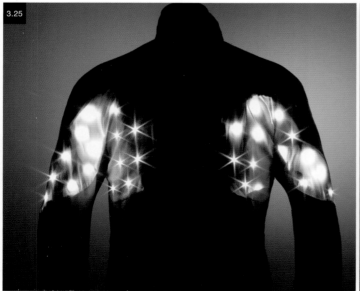

3.27

The Under Armour E39 monitors athletes' vital signs and intensity levels, sending the information wirelessly to a separate device, where third parties can monitor them.

released the E39 for use during the National Football League (NFL) Scouting Combine, a week-long showcase that occurs every February, where college football players perform physical and mental tests in front of NFL coaches, general managers, and scouts in preparation for the NFL draft.

Vital signs

A smart compression top, the E39 has a built-in body-monitoring device that sits just below an athlete's sternum. Outfitted with the E39, each athlete had his vital signs (including heart rate, and respiratory and intensity levels) recorded and transmitted to the Bluetooth-enabled smartphones, tablets, or laptops of the observing NFL staffers. Along with their size, speed, agility, vertical jumping height, and endurance, coaches were able to assess how much energy each athlete was exerting during the drills. Were they giving it their all, or was there more room to push for greater performance?

Along with endurance and fitness assessment, the E39 can also analyze body movement and mechanics. Whether assessing a player for the draft or while they are out on the field, feedback from the device allows trainers to see exactly how a runner accelerates. By measuring acceleration and

3.27

3.28

3.28
The Reebok Checklight by MC10 is a
device that is worn under an athlete's
helmet during contact sports. Using
an accelerometer and a gyroscope the
Checklight can detect impact damages
to the head. If those damages are
dangerous, the Checklight turns on an
LED signal.

direction on separate sides of the body, trainers can then
analyze a player's stride to see if it can be optimized for
speed by bringing the two sides into better synchronization.
Poor synchronization can create drag and reduce speed.

The E39 can even be used to minimize the risk of strain
or injury by alerting medical staff to the warning signs of a
player who is not operating at full capacity, allowing them
to recommend removing them from a game or practice drill.
Early signs of dehydration and excess body heat can also
be detected and averted. With more than 50 colleges using
the E39, and even greater numbers of younger athletes
participating in football, sports medicine experts see these
metrics as a positive addition to game safety.

Preventing injury was also the driving factor behind the
Smart Pitcher shirt developed by three engineering students
at Northeastern University in Boston, Massachusetts.
Previously, in order to analyze the biomechanics of his
fastball, a pitcher had to visit a lab; this compression
shirt—outfitted with accelerometers and motion sensors on
the back, biceps, and forearm—can now transmit this data
from the pitcher's mound. Major league pitcher injuries cost
teams many millions of dollars in lost salaries, not to mention
player injury and recovery time. This shirt delivers information
in real time as a player of any level is pitching. Beyond injury
prevention, the Smart Pitcher shirt could also be used for
instruction, virtual coaching, and even online and remote
coaching and recruitment for young players. The creators
believe that the shirt could easily be adapted for use in other
sports such as tennis and basketball.

Protection

3.29
DuPont's Nomex® fabrics are used across a wide variety of professions to protect against heat and flame. Here, firefighters battle a blaze in close proximity to flames; they are protected from burns by their clothing made from Nomex® fabrics.

Smart textiles offer more protection than ever before, and with less weight and greater flexibility. In the military, police, firefighting, and manufacturing sector, these materials form barriers to impact, flames, lasers, bullets, knives, gases, and radiation, which would otherwise be impossible to survive without injury. Creating a portable environment, protective clothing has made navigating these harsh conditions not only possible, but safer and easier, leading to improved performance and safety for both workers and the public. From business travelers, schoolteachers, and politicians, to people whose work brings them face to face with the threat of terrorism, this new class of materials is bringing protective qualities to everyday clothing that were once only offered to elite personnel.

Bulletproof vests have been a staple of military and law-enforcement personnel for over a century. Many innovations over the years—including the creation of Kevlar® and other similar materials—have led to safer, lighter, more comfortable vests. As part of a military uniform these bulky vests, which rely on multiple layers of fabric to function, are not an issue, but under business clothing they are awkward and uncomfortable. Now a new material made with carbon nanotubes is being used to create soft, more fashion-forward clothing with built-in protection from bullets and knife blades.

3.29

For a business suit that can literally save your life, Michael Nguyen's Toronto-based tailor shop, Garrison Bespoke, has developed a bulletproof, made-to-measure three-piece suit by adapting a fabric first developed for the military. Many professions, from politics and international finance to oil and diamonds, expose people to dangers and potentially life-threatening situations. These clients are looking for professional clothing that is protective, discreet, and good-looking. Layers of this newly patented, carbon nano-fiber suiting, which is 50 percent lighter and much more flexible than Kevlar®, are incorporated into the traditional padding of a men's tailored vest, jacket, and pants. The custom-made $20,000 suits have been tested to stop rounds from 9mm, .22, and .45 caliber handguns.

3.30
The bulletproof suits are tailored in the Garrison Bespoke shop run by Michael Nguyen. Each suit contains layers of carbon nano-fiber, a lightweight bullet-proof material.

3.31
The suit has been proven to stop rounds from 9mm, .22 and .45 caliber handguns. A useful garment for anyone who needs to be safe and sleek.

3.32
The Hövding airbag helmet disguises itself as a stylish scarf to appeal to people's general fashion aesthetics while keeping them safe on their bicycles (see image 3.33 on page 100 for impact shot).

The Armored Armani

Since the 2012 Sandy Hook Elementary School shooting in Connecticut, manufacturers of bulletproof clothing and body armor have seen unprecedented sales in everything from jackets and windbreakers to T-shirts, corsets, and even wetsuits. In Bogotá, Colombia, Miguel Caballero (nicknamed the Armored Armani) makes lightweight, fashion-forward bulletproof gear for men, women, and children. Sports jackets and overcoats, motorcycle jackets, casual clothing for both warm and cold climates, and even a bulletproof tie are made to protect everyone from UN peacekeepers to VIPs and rap artists. The demand for fashionable off-the-rack protective clothing is a growing business. There are now Miguel Caballero fashion boutiques in Mexico City, Guatemala City, Johannesburg, and London.

From bullets to bicycles, life-saving protective gear that doubles as everyday clothing can save lives while also reducing people's reluctance to wear it. Hövding is a revolutionary new bicycle helmet that is not only the safest helmet made today, it is also the most discreet. Unlike a traditional helmet, it is worn as a collar around your neck.

The product started life as a master's thesis by Anna Haupt and Terese Alstin, while studying industrial design at the University of Lund in Sweden. The collar contains a uniquely designed airbag that inflates on impact to cover the entire head; it protects from rear, side, and front impacts without obstructing the wearer's field of vision. The bag is protected by an abrasion-resistant lightweight nylon, and interchangeable covers allow the Hövding to coordinate with your wardrobe, so it can be worn to look more like a scarf than a helmet.

Industrial workers in industries such as mining, deep-sea fishing, steel manufacturing, oil and nuclear energy, and many others, rely on protective clothing to be able to do their jobs, and to protect them from adverse conditions and life-threatening accidents. With so much money invested in these global industries, research and innovations in new materials to meet their demands has been well funded.

Lasers are used for a growing number of operations in today's industrial workplace, including cutting, alignment, annealing, drilling, dynamic balancing, metrology, sealing, soldering, and lithography—to name just a few. Despite

3.32

3.33

the growing use of hand-guided lasers, though, until
recently the only classified and certified protective gear
available for workers was eyewear. Many operators work
in close proximity to these lasers, where a small error can
cause severe burns, while infrared radiation can penetrate
deeper into the body, causing harm to blood vessels and
surrounding tissue. Soon, workers will be able to protect their
bodies and hands as well as their eyes.

The German company Laser Zentrum Hannover (LZH)
has been working on a European Commission-funded
project to develop a full range of protective clothing for laser
operators. The collection, which includes lab coats, aprons,
pants, and gloves is designed with both a passive and active
protection system. Multiple layers of fabric work in tandem to
first reflect as much of the laser as possible, then to absorb
and disperse the remaining radiation to mitigate the damage
it might inflict. The protective layers are also embedded with

3.33
Upon impact the helmet releases and
inflates, covering all but the front of
the wearer's head and protecting it
from impact.

3.34

These gloves protect the hands of laser operators. Some of the gloves' multiple layers reflect the laser beam while others act to absorb the radiation.

3.35

Using a multilayer design, the Hydro jacket repels extreme heat, preventing it from penetrating the interior of the jacket. The lining not only keeps the wearer cool but uses hydrogel padding to manage moisture.

sensors which create an active monitoring system that will turn off the laser when it detects radiation. The combination of a physical barrier and an active smart-sensing system brings unparalleled safety to this fast-growing industry.

Barrier clothing

Barrier clothing has been used in both fire protection and arctic sea gear. The Hydro jacket is a fireman's kit designed by Grado Zero Espace that uses a thermal- and moisture-management base layer created by a super-water-absorbent polymer coating. It is designed to give greater comfort and safety to firefighters. Like the laser protective clothing, the Hydro jacket is built with a multilayered fabric system. It has an external layer that reflects sun rays, a fire-resistant coating that is activated by the heat of a fire, and a second layer that works as a thermal barrier and dissipates heat buildup. Safe Hydrogel Padding is used in the lining to create the best moisture-management system available, while remaining incredibly lightweight and flexible.

Hydrogels are hydrophilic polymer networks that are able to retain large quantities of water while maintaining their structure. They don't dissolve in water, instead swelling to

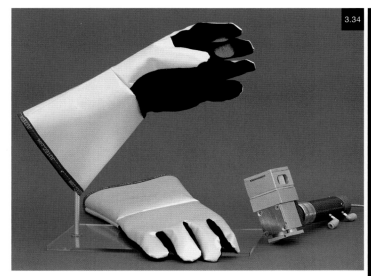

3.34

3.35

Hydro*Jacket

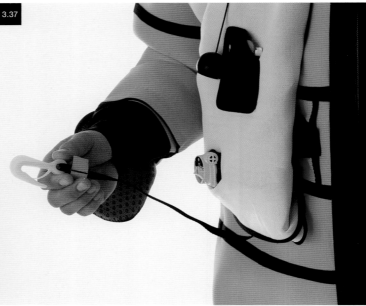

hold the increase in water volume without breaking. This super-absorbent property has led to their use in products across many industries, such as cosmetics, pharmaceuticals, artificial organs and tissue engineering, wound dressings, contact lenses, fire protection, and most notably, disposable baby diapers. The Hydro jacket uses this technology in a coating and an insulating capacity. The coating on the base layer absorbs and retains the firefighter's perspiration to keep him/her as dry and cool as possible; in the second layer it works as a heat dissipator.

Survival at sea

Emergency survival suits for use on board merchant ships, fishing vessels, and offshore installations have to withstand extreme conditions of near-freezing water, puncture, abrasion, and even fire. Survival at sea depends on this piece of equipment working perfectly—if just one component of a survival suit breaks down, the rate of survival is dramatically decreased. A combination of design features, fabric, and construction must work together to create a self-contained environment that protects the sailor from exposure to an accident, and ultimately the ocean, until s/he can be rescued.

The innovative Sea Arctic immersion suit by Hansen is a fully integrated all-in-one survival suit that is built to withstand years of use. It is made from extra-durable, flame-

3.36
The Sea Arctic suit is tailored to the needs of the offshore worker. Using a multitude of systems, the suit is an important protective component during everyday operations and a lifesaver in the event of an accident.

3.37
The reliability of the functionality of the Sea Arctic suit is integral to the survival rate of the wearer. A person can survive for up to six hours submerged in arctic water while wearing it.

resistant neoprene, which is also designed to withstand ageing. The suit's fluorescent dye is formulated to retain its color when wet to maintain visibility when in the water, and its seams are sealed so that the suit retains buoyancy. It also has a built-in self-righting device in case the wearer is swept overboard in rough seas. Construction details include a mitten hand cover that is permanently attached with elastic, and a plastic hood cover that can completely enclose the suit. Amazingly, the suit has been approved for up to six hours' immersion in arctic waters.

Integrated environmental monitoring system

Mining accidents claim thousands of lives every year. The Mobile Monitoring Station, a new smart jacket for miners with an integrated environmental monitoring system, could save many of those lives. The jacket has sensors that monitor the environmental and physical conditions underground, and will alert the miner to any changes while also transmitting data to the mining company in real time to help prevent any accidents before they happen. The sensors will monitor the wearer's vital signs, too, looking for elevated heart rates, and early warning signs of silicosis (a lung disease caused by breathing in fine dust particles) and hearing loss from excess noise. The sensors will also monitor the atmosphere for radon (a radioactive, odorless gas which is also a known carcinogen), welding fumes, and other toxic metals; this can be customized depending on what is being mined. The jacket alerts its wearer to dangerous levels of these and other elements, and the data is simultaneously transmitted wirelessly to the surface, where software will sound an alarm when an unsafe level is reached in the mine.

Mining is a dangerous job, but with the help of smart technology workers will be able to monitor their own safety while companies work toward creating the safest working environment possible.

Sustainability: Renewable Energy and Energy Conservation

Fabric batteries? Kicks that generate electricity? Solar energy-generating clothing? Scrunchable antennae? All of these things sound like they are the product of an overactive imagination, but they are all in fact real products. Researchers are at work in laboratories around the world on projects that range from a device that will charge your cell phone as you run, to data that can be stored in your pocket, literally. Perhaps the most exciting aspect of this research, though, is its potential impact on sustainability, with proposals to generate energy from human kinetics and environmental conditions such as sun, wind, and sound.

Researchers have now created a stretchable lithium-ion battery that is able to extend to four times its initial length with full recovery. This innovative product is completely flexible and continues working when it is folded, twisted, or stretched, making it the building block for solar power-generating clothes. Developed by an international team of engineers from China, Korea, and Illinois, the group see the SD (secure digital) card-sized battery being used in clothing (particularly activewear) that generates power, a flexible touch-sensitive skin for robots, temporary tattoos that monitor your vital signs, and many other flexible futuristic wearable applications.

Even old-school technology can be adapted to generate new ideas and results. At the University of Auckland in New Zealand, a group of researchers have adapted a simple, inexpensive rubber generator for a pair of shoes that can be

3.38
A lithium-ion battery that is capable of stretching up to four times its size. The battery can maintain operation under stress in many different directions.

3.39
This battery is a leading step in the development of solar-powered clothing due to its stretch and flexibility.

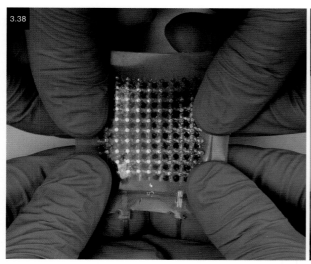
3.38

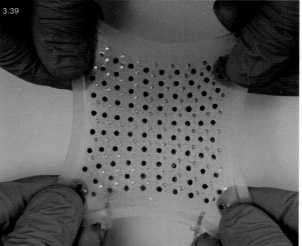
3.39

3.40

The Biostamp temporary tattoo by MC10 powers itself through heat changes from the body. The stamp can harvest enough energy to power both stationary and hand-held medical devices as well as gather data from the wearer to help inform better, faster, and more accurate healthcare decisions.

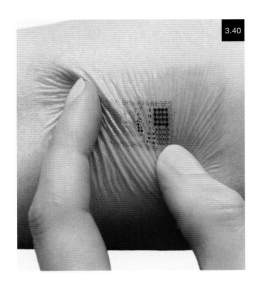

3.40

used to harvest energy from walking. Over time, the movement of walking builds up to a watt of power—the amount used by your cell phone. The generator is made from artificial muscle material consisting of dielectric elastomer actuators (smart material systems that produce large strains), which compress and expand gradually, forming the charge. Each generator is estimated to cost under $4 to manufacture and would be an easy add-on for any number of clothing or shoe designs.

Harvesting energy

Ever since *Invasion of the Body Snatchers*, the idea of harvesting energy from the human body has fascinated science-fiction writers and scientists alike. Advances in nanotechnology may now turn this fantasy into a reality. Thermoelectric conversion materials can convert changes in temperature into electricity, which can then be stored in a battery, or used directly to power or charge a device. Scientists from Fujifilm, in collaboration with Japan's National Institute of Advanced Industrial Science and Technology (AIST), have developed a new highly flexible thermoelectric conversion material that can be printed onto the surface of textiles, or directly onto the skin as a temporary tattoo. It can detect and harvest electricity in minute changes in temperature, as small as 2°F (-1°C). The researchers anticipate a number of applications for their product—it could be used by patients to power their own medical devices, on clothing to charge batteries for hand-held devices, and on solar panels to increase their efficiency. A piezoelectric material is one that can generate a voltage when pressure is applied through movement, bending, or vibration. To harvest energy from the environment, students at the American University of Sharjah in the United Arab Emirates have been looking for places where a lot of mechanical energy and noise is wasted, such as sports stadiums, footsteps on sidewalks, and busy highways. They wonder if sound waves, especially in loud environments, could create enough pressure to generate sufficient current to power a portable device.

Other groups of researchers are also working toward new applications for piezoelectric materials. These could include fabrics that charge our cell phones as we talk or listen to music; athletic clothing and shoes that generate and store electricity as we exercise; noise-reducing interior fabrics that could in turn power lighting and small appliances; and even architectural fabrics that could reduce the noise impact of traffic and, in turn, power street lights. All of these applications are in development, not far from the market.

Built-in communicators, like something from a sci-fi movie, could also become a reality in the near future, with the invention of a stretchable, flexible fabric antenna. Finnish company Patria have developed a wearable antenna that can not only conduct electricity, and harvest and store energy and data, it can also send and receive satellite communication signals with GPS location data. Eventually the patch will allow for incoming calls as well. Just like Captain Kirk, you'll be able to tap your shirt for an incoming call.

Currently, the antenna is able to send a signal using the Cospas-Sarsat worldwide search-and-rescue satellite system. Backcountry skiers and other adventurers generally carry an emergency beacon with them, but at the risk of it falling out of a pocket, or being lost if the wearer becomes separated from their team. This antenna could be sewn right into clothing, ensuring certain rescue in the event of an accident.

No-wash clothing

Water conservation is today a high priority, and globally we need to reduce water consumption. Around 780 million people lack access to clean drinking water; in the West an estimated 22 percent of household water is used for laundry alone.

Leading a group of researchers in China, Mingce Long of Shanghai Jiao Tong University, and Deyong Wu of the Hubei University for Nationalities, are working on developing clothes that clean themselves with sunlight and no water. The team drew on previous work that discovered a titanium oxide solution able to remove stains in UV light. Since the use of UV light is not very practical, they turned their attention to sunlight instead. They developed a nano-particle solution of titanium oxide in liquid form, into which they could then dip a cotton swatch. Once the swatch had been pressed and dried, the cotton was treated with a coating of silver iodide to enhance its light sensitivity. They stained the fabric with orange dye and exposed it to sunlight. The dye was completely broken down and, when tested, the cotton surface was free of bacteria. The team hope their discovery will lead to clothing that never needs to be washed, making laundry detergent a thing of the past.

3.41
The development of clothes that self-clean when they are exposed to sunlight could save the water that would otherwise be used to clean them.

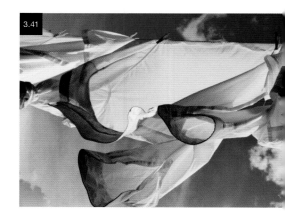

3.41

Medical: Prevention and Healing

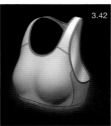

3.42

3.42
First Warning Systems has developed a bra that increases the privacy, efficiency, and accuracy of screening breast tissue for cancer. When worn for 12 hours, the bra can detect subtle temperature and tissue changes in the breast which are more accurate indicators of a tumor than the traditional mammogram.

It's only natural to want our clothing to serve multiple purposes. We want it to look good, feel good, and now we want it to have added benefits for our health and well-being. The field of cosmeto-textiles is one of the fastest-growing areas of smart textiles. This group of fabrics uses nanotechnology to embed everything from moisturizers and perfumes to anti-ageing and slimming agents into fabrics. French and Italian companies lead this industry, and the applications are primarily in womenswear and lingerie.

Beyond cosmetics, smart textiles are revolutionizing medicine in diagnosis, wound care, and medical replacement. A new bra has the ability to detect tumors before both self breast exams and mammograms. It looks like a sports bra but the cups are lined with a series of 16 small sensors that read temperature changes deep in the breast tissue. A patient takes the bra home and wears it for 12 hours; all the while the device is storing the data which is then downloaded and analyzed with pattern-recognition software to see if anything is amiss. In early testing the bra is more accurate than the current methods of testing for breast cancer, and far less invasive.

Another innovation is the Bellyband Prenatal Monitor, still in the development stage at Drexel University's ExCITe Center in Philadelphia, Pennsylvania. The first prototypes were created by the ExCITe Center's research team, together with researchers at the school's College of Engineering and School of Biomedical Engineering, Science, and Health Systems. This is a revolutionary machine-knit wireless monitor that the expectant mother wears during labor to monitor the baby's heart rate and stress levels while *in utero*. The soft, flexible knitted band is warm and comforting. It is also designed to be freeing and soothing for the mother during the stressful birthing process. Conductive yarns are knitted into the band; these are connected to sensors that monitor the baby through the mother's abdomen. Wireless transmission leaves the mother free from any restricting wires, while soft yarns stretch around her belly as she moves during the birth.

Smart textiles are also used in many other areas of the medical field, from surgical gowns and wound dressings to implants, and innovations are constantly being introduced. Smart wound dressings that have medications embedded in their fabric can heal wounds faster and with less risk of

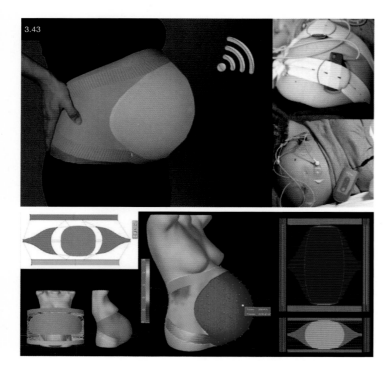

3.43
The knitted Bellyband (top left-hand frame), developed by a research team at Drexel University in Pennsylvania, uses conductive yarns and sensors to deliver information during childbirth wirelessly, allowing the mother to move freely without connection to traditional monitoring devices (top right-hand frames). The lower frames show the Bellyband development process. To develop the knitted pattern, the designers used a virtual stitching process simulated by computer-generated imagery.

infection. There are also bandages being designed with special glass and bamboo fibers, which are proven to make blood clot faster, and with a release feature that prevents them sticking to the clot once it is formed. Some bandages are designed with embedded super-absorbent polymers so that they stay dry longer, while others have antimicrobials embedded to resist infection.

There are even bandages programmed to change color when they detect a rise in skin temperature or a change in pH level—an indication of possible infection—so as to alert the attending physician. This saves valuable healing time, especially with time-sensitive wounds like burns.

The Dattoo

The Dattoo (DNA tattoo) is a concept for a temporary tattoo that will be used not too far into the future as both a diagnostic tool and an interface for the body. It will be able to read blood-sugar levels, heart rate, pH, and any other measurable data conceivable. The temporary tattoo will be printed with the wearer's specific DNA as an identifier, and will be able to be washed off when no longer needed. The display will be an OLED (organic light-emitting diode) monitor with a dynamic touchscreen interface, and the whole device will be powered by the wearer's own body heat.

3.44

Using smart textile-enhanced medical equipment such as super-absorbent polymer-embedded bandages can help wounds heal faster and more effectively.

Closer to actual production, the Future Force Warrior project is a military uniform that protects, communicates, and diagnoses injury—a fully integrated twenty-first-century infantryman combat system. It's part of a larger project that the US military has developed over the past decade that includes the Land Warrior, the Soldier Integrated Protective Ensemble, and the Transformation of the US Army projects. If a soldier is injured in battle, the fabric in the uniform will detect his/her vital signs and bullet wounds, determining the estimated depth of penetration and the effect on surrounding organs. Fabric sensors will also detect exposure to chemical, biological, radiological, nuclear, and explosive agents. It will be able to do this by tapping into biomarkers found in the soldier's blood, saliva, perspiration, and urine.

These functions are in addition to the uniform's state-of-the-art body armor, which is lightweight and designed to disperse the impact of a bullet over a wider area to decrease injuries such as broken ribs. A soldier will be able to chat online with a tactical local and wide-area network, using an onboard computer located at the base of his/her back. The wiring of the computer runs through the uniform and the soldier's helmet, which is wired with sensors that detect cranial vibration, rendering the need for a microphone obsolete. Each soldier will have the ability to share voice, data, and video in real time with vehicles, aircraft, and other soldiers.

The onboard computer will also track the overall physiological picture of the soldier's performance, including body core temperature, skin temperature, heart rate, whether s/he is standing or prone, and how much water s/he has drunk. A medic may be miles away, but with the data collected by the uniform and helmet, the soldier can be treated remotely, preventing stress-related injuries and gaining precious time.

The Future Force Warrior project combines smart fabrics and wearable technology to create the most advanced wearable communications systems in military uniform design today.

PEOPLE AND PROCESS

Have you ever wanted to peer into the window of an accomplished designer's studio to see just what was going on? Perhaps join their team for a week or two, work on a project to gain some insight into their process, and take away a tip or trick to use when investigating a design solution of your own? Sharing the process of others is perhaps the greatest learning tool available to designers. How we solve problems and look at materials, generate concepts, and search for solutions is unique to each of us but there are commonalities. Insight, ideation, visual thinking, creation, prototyping, testing, and narrative are tools all designers share, but each discipline emphasizes different aspects and executes them in a different manner.

Generally speaking, each design discipline has its own approach to problem-solving and the creative process. A fashion designer and a fine artist will approach a project in completely different ways, each using a different perspective to reach a solution. When working with new materials and emerging technology, it can be an advantage to look at how practitioners in other disciplines create their work. Traditional methods of working may have to be completely rethought. Working with emerging technologies may require mastering new skills or acquiring knowledge in fields with which you are unfamiliar.

In the following chapter, we will look at the creative process of five disciplines: industrial design, fine art, fashion, architecture, and engineering. Each takes a different approach to conceptualizing, creating, and working with smart materials.

Case studies in each of the disciplines can teach us specifics about how each field applies its practice to working with smart textiles and, in many cases, wearable technology.

You will see that many designers advocate collaboration and stress the need for working with others who have expertise in other areas. Multidisciplinary is the current buzzword, and for good reason. Teams of designers, engineers, and artists bring different skills and approaches to the studio, all feeding off each other to revolutionize and retool product design, architecture, and apparel for the future. Looking into the studios of the featured designers, fine artists, architects, and engineers, we will see first-hand how they have approached working with smart textiles, creating a roadmap for you to apply to your own practice.

Industrial Design—Driven By the User

Industrial designers are focused on the end-use of a product—who will use it and why. They have been known to go to great lengths to understand the use of their products, and have at times extended their reach to rethink what it actually means to "use" a product, redesigning the whole experience from the ground up by rethinking an entire system or redefining an entire product category in the process of designing a single product. This focus on everyday activities and problems faced by the end-user is what drives the process of the industrial designer. To use this process successfully you have to empathize with the customer, putting yourself in their position. The industrial design approach is, therefore, user-driven.

Many large product design firms have multidisciplinary teams of designers, anthropologists, and researchers who collaborate during the design process to establish the criteria that will drive a design solution. The driving factor of this process is observation, often enhanced by personal experience, but grounded in empathy with the person who will be wearing the garment or interacting with the product. The term "interaction design" is often used interchangeably with "industrial design," because it is the interaction between user and object that the designer focuses on when searching for a solution.

This is one reason why many active sportswear companies, for example, are founded by athletes or recruit the input of elite-level athletes and continue to develop a culture of sports lifestyle in their product development and sales staff. Having participated in a particular sport, especially at a high level of competition, grants insight into the unmet needs of other athletes.

Summary

The focus on the end-user drives the creative cycle from start to finish for those engaged in the industrial design process. The designers in both case studies that follow were led to develop proprietary smart fabrics that met the needs of their customer. The unwavering belief that the user experience was the most important element of the design was the driver behind every design and production decision. Without wanting to downplay the importance of aesthetics—as evidenced by CuteCircuit's design of their electronic circuits, aesthetics are extremely important—but ultimately it is *how the user feels and the experience they have using the products* that drive the design process.

113

Michele Stinco

PolychromeLab

Michele Stinco uses the user-driven approach in designing technical outerwear for his company Polychromelab in Switzerland. An avid mountaineer, he was frustrated by the fact that he was never dressed perfectly for the dynamic temperature changes in an alpine climate. Sometimes he was too hot, sometimes not. The weather forecast was often inaccurate, and you could start the day at 65°F (18°C) and by the time you arrived at the peak, it could be in the low twenties, with wind chill, making it feel far colder. Stinco explains, "During the day, whether you're in the mountains or in the city, you experience many weather changes. The big companies were always saying that something new was on the way that would meet my needs, but the reality was that every time it was just a small change in design or fabric weight."

Driven by his need to find a better jacket, Stinco, who is a trained activewear designer, started with a list of the properties he needed his jacket to have. It had to keep him dry, be highly breathable, water resistant, stretchable, and abrasion resistant. It also needed to adapt to any weather conditions by absorbing warmth when it was cool, and reflecting heat when it was warm. Once the problem was defined, his next step was to embark on a "feasibility study." Stinco explains, "We tested existing products and fabrics with different weather applications. After a month of testing and talking with fabric suppliers, I found the information I needed to create something that would be able to fulfill all my needs."

His solution was to create a three-layer modular fabric. "The basic problem with many three-layer fabrics from other suppliers is that they are more into look, feel, and water resistance, and are not into the textiles of physics. The most challenging part of the process was finding the right components for my textile's two-face fabrics and membrane, then figuring out how to laminate them together, making the components UV/IR (ultraviolet and infrared) absorbing and UV/IR reflecting."

Sketching designs

Once the fabric was created, Stinco began sketching designs for his Alta Quota jacket. The sketches were refined, and, after the detailed technical drawings had been completed, his next step was to find a supplier that could prototype and manufacture his jacket. After many trials in Austria and Germany, he moved to Italy, where he found a manufacturer to complete the project.

There were a number of reasons for this decision. Stinco explains, "I am Italian and in Italy you have many possibilities of doing something new and different. In Austria and Germany, the process is very rigid, not especially creative but the work is good quality. In Italy life is chaotic and difficult to manage, but this flexibility makes a difference when working with innovative products." However, he goes on to explain that "the scientific world is in Austria, with the Institute of Sports Science in Innsbruck and the Research

4.1

4.2

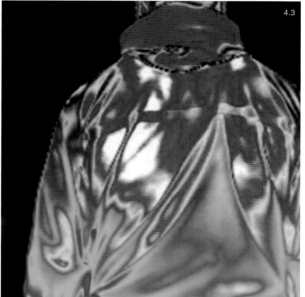

4.3

4.4

4.1

The Polychromelab Alta Quota reversible jacket is the first lightweight single-layer jacket suitable for the extreme temperature changes experienced in mountaineering: between base camp and summit these can vary by as much as 50°F (10°C).

4.2

Here the textile's breathability and the jacket's ability to respond to an increase in the user's body temperature are tested on a treadmill.

4.3

Thermal imaging was used when the team designed the Roccia Rossa Bionica jacket, which won the 2014/2015 ISPO Award. The blue side of the jacket absorbs 100 percent of UV radiation and reflects 45 percent of the IR [explain] to keep the adventurer warmer during rest periods and in colder weather.

4,.4

Fabrics developed by Polychromelab are tested at the company's Textile Research Lab at Mount Glungezer, Tyrol, where winds reach 264 kmh (164 mph) and test dummies have to be lashed in place.

Institute for Textile Chemistry and Textile Physics in Dornbirn, which includes our own scientific lab." Stinco therefore chose to combine the best of both worlds by prototyping in Italy and testing in Austria. The entire process of developing the fabric and a working prototype took two years.

Beyond fabric

There are some innovations in the construction of the jacket beyond the fabric. The Alta Quota is unique in that it is completely reversible: One side is black and the other reflective silver. A challenging construction issue that the team faced was applying the tape for the seam sealing to the reflective silver side. All the jacket's details needed to be executed to perfection since everything is visible—the ergonomically designed pockets, the zipper quality, and other details. Finally, the jacket was designed with an awareness of its carbon footprint: The garment involves travel of no more than 500 miles (800km) from the manufacturer of the fabric to the store's shelves. Every detail, from fabric development to how and where it is

constructed, was considered in the creation of the product and its story.

Secrecy is paramount for Polychromelab. The fabrics they have developed are patented and took years of prototyping and testing to perfect. The most difficult step in the design process for Stinco was finding suppliers and manufacturers that he could trust. He admits that in the early days a German patternmaker dealt a near-fatal blow to the company when he failed to maintain the confidentiality of the designs. But now, with Italian suppliers secured and the launch of a second technical fabric, Polychromelab has established itself as a premier technical outerwear resource.

Testing was key to proving the functionality of both the fabric and the design. Stinco and his team tested the Alta Quota jacket in many different environments and in a variety of different conditions. Scientific testing was carried out both in the lab and on the mountainside.

When reflecting back on his design process for the Alta Quota, Stinco says, "The most important point to understand is that the complete project was the problem. To solve the problem, we had to compose a three-layer fabric, design a jacket, find trustworthy suppliers, search for the correct components, find the money to finance the project, and create the correct communication so that people would understand that we were talking about textile physics and not just a jacket that is silver and black. So you see, my one problem led us to solve many problems."

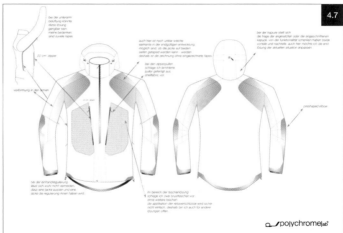

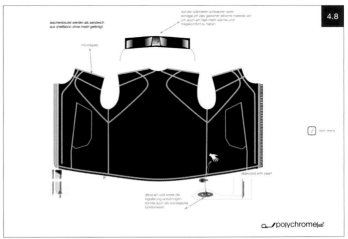

4.5

Designer Michele Stinco, the founder of Polychromelab, and Elisabeth Frey, the managing director, discuss the development process of the performance garment.

4.6

A team of highly trained technical designers work to fit, test, and manufacture the jackets to the highest standards. The Alta Quota jacket took over two years to develop before it was ready for distribution.

4.7 & 4.8

Technical design sketches depicting both the lighter layer's style lines and the darker layer's construction. To create a perfectly reversible and functional garment, the team needed to develop its own manufacturing methods for seam finishing and garment closures.

4.9

Each side of the Alta Quota jacket works with the outdoor temperature to help control body temperature. The dark-colored matt side is designed to absorb heat in cold weather; when the outdoor temperature gets warmer, the jacket can be reversed and the light-colored side reflects the sun and UV rays to cool the body.

4.10

Francesca Rosella

CuteCircuit

Francesca Rosella, one half of the design duo CuteCircuit, has pioneered some of the most recognizable wearable technology and e-textile garments available today. CuteCircuit are based in London's Shoreditch area, and together with partner Ryan Genz, Rosella has worked on interactive garments for performer clients and for couture and ready-to-wear collections. Their signature achievement and ongoing project, however, is The Hug Shirt™.

Rosella and Genz were looking for a wearable technology project that would bring people together; something that would humanize technology and embody a simple human need—touch. Rosella explains, "We talked to people about things they like and the message we received is that people really need to be hugged; there is a need for a simple human touch. We thought about how to embody this touch feeling. Hugging is a physical sensation, so we decided to design a garment that hugs you."

Working with test subjects

Approaching the project from the viewpoint of the end-user, the pair worked with a group of test subjects to research and analyze what it means to be hugged. Rosella recalls, "We went through a very, very long process. First we started by giving people different materials to touch and we would ask them what they liked about the feel and touch of each material. We gave them little plastic balls, squishy little things, like little plush toys." Finding that different textures and materials would generate different reactions, they were able to determine some common elements that people associate with the warmth of a hug. They then used this information to guide them when searching for the fabrics that would be used for the shirt design.

The designers then followed the materials inquiry with more research. They assembled 50 people in a room and gave them all white T-shirts to wear. Rosella explains, "It became a huge body-storming session (the physical equivalent of a brainstorming session), where we asked these people to hug each other for a very, very long time, while Ryan and I went around with red markers recording the positions of their hands on the other person's body." This research became the primary driver behind the placement and positioning of the "hug" sensors and actuators in future prototypes.

"The current design has white and red areas. The red areas are the places that people touched the most when they hugged someone," explains Rosella. "It's like a really interesting physical way of designing that is not just like, oh, we got this idea and then we're going to make this normal fashion design. No, it's more like asking people what would you like to wear, and then when you hear lots of different answers, you apply the best ones. It's almost as if everybody is designing together. We find this process more emotionally fulfilling. We try to go through this process to give

4.11

4.12

SMART FABRICS AND INTERACTIVE TEXTILES

4.13

4.10 (page 118)
The Hug Shirt™, by CuteCircuit, was invented in 2002 and is the world's first haptic (touch) telecommunications device. You can send a hug through the shirt to distant loved ones.

4.11
The CuteCircuit Catsuit was designed for Katy Perry in 2011, when she appeared on *American Idol*. It is decorated with interactive luminous crystals.

4.12
The Galaxy Dress is the centerpiece of the "Fast Forward: Inventing the Future" exhibit at the Museum of Science and Industry in Chicago. Extra-thin full-color pixels are embroidered onto stretch silk so that the design follows the contours of the body and acts as close to normal fabric as possible.

4.13
Four black-and-white leather jackets with over 5,000 pixels of light were designed for legendary rock band U2's 360° Tour. Each jacket was synchronized and capable of displaying any number of unique designs and images.

something to people that is both fulfilling and emotional—to us and to them."

Rosella and Genz started developing their design concept of connecting people while they were both researchers at the Interaction Design Institute Ivrea, a graduate school near Turin in Italy focusing on interactive design, which has now merged with Domus Academy, Milan. Rosella was convinced that fashion and technology could merge to create what she calls Human–Human Interaction (HHI), the evolution of Human–Computer Interaction (HCI).

"We want to cut out the middleman, the computer, and just let people connect with each other. In the future, the computer is going to disappear and the interface is going to be us. If you touch a surface somewhere, information pops up, or if you stand in a room, others will know who you are because their garment interface told them. It's just a sense of better connectedness. You see, some people think wearable technology is going to replicate whatever interface they already have. A keyboard will be hiding somewhere and a monitor hiding somewhere else. That's not the future. We don't need to replicate something that already exists. We're going to eliminate all of these other things and have a new kind of interface that is much more intuitive."

This drive to connect people is also seen in CuteCircuit's work with stage performers. In 2011, Italian singer Laura Pausini approached CuteCircuit to design a costume for an upcoming world tour.

She wanted a really amazing costume to wear while performing her hit song "Invece No," which would also complement the incredible set that had been designed by British architect Mark Fisher, best known for his work with U2, Pink Floyd, and Cirque du Soleil. Rosella describes the stage as "this massive set that had a main stage and a 'B' stage with people in between in a mosh pit. The plan was for [Pausini] to fly over these people."

Approaching the design problem from the audience's viewpoint, Rosella's goal was to create something magical. She wanted the audience to really be immersed in the environment and give them something that they had never felt before. She continues, "Performers are looking for a sense of connecting with their audience on a deeper level. It's not just about delivering the songs, but giving people an experience so that they feel like they are participating. It's not just the singer having fun, but everyone having fun with them."

The design that they created was a dress with a 14 ¾ft (4.5m)-long skirt, embroidered with LED lights. The designers used over 165ft (50m) of silk to create the dress, so that when Pausini flew over the people in the pit, the hem of her skirt would actually touch them. Hoisted above the crowd on a huge swing, she would move her legs to ripple the chiffon. As the fabric flowed, the embedded LED lights would start to come up and sparkle. Rosella remembers the excitement of seeing her up there looking like "a giant jellyfish"!

Once conceived, CuteCircuit's intricate designs take a lot of technical knowledge to execute, and most of this knowledge came from years of trial and error. The first dress they created, *The Galaxy Dress*—now part of the permanent collection of the Museum of Science and Industry in Chicago—was their first attempt at using a fabric that was covered with tiny LEDs. The entire dress had conductive fabric and LEDs hand-embroidered into it, and took six months to complete.

LED-infused fabric

Soon after that, Rosella and Genz decided to develop their own fabric, one that would make working with lights and circuits simpler and faster. Their solution was a dual-layered fabric consisting of LED-infused fabric modules that plug into a base layer. The modules can be arranged to interpret many different designs, and house both the LEDs and what Rosella calls "the brain," a mini-circuit that controls the programming for the lights on that module.

Rosella explains, "The brain is where the intelligence is. For example, the garment has many of these modules, each with tiny little connected brains that all talk to the large fabric screen." The brains are small sequences that are triggered by a larger program running through the base fabric. Their fabric, which doesn't have wires running through it, can be cut and connected to fit the body. It took CuteCircuit over ten years to perfect their patented fabric.

4.14

In her studio at CuteCircuit, Francesca Rosella creates a design sketch for each project. From there, she and her partner Ryan Genz work on how to incorporate the electronics seamlessly into the design.

4.15

Beading detail of a dress being created in the studio. Each of these jewels will ultimately be lit with LEDs and controlled by custom-designed software that is activated from a smartphone or hand-held device.

4.16

CuteCircuit's *Hug Shirt*™ is embedded with sensors that feel the duration, strength, and location of the touches it receives and transmits that information to actuators in a partner shirt.

4.17

CUTECIRCUIT 4.18

4.17
Designed for singer Laura Pausini, this dress has a 14¾-foot- (4.5-meter-) long illuminated skirt. It is lit up by the 5,670 luminous pixels which shine from top to bottom through the delicate aqua techno-tulle, creating different patterns which synchronize with Pausini's music.

4.18
Designer Francesca Rosella and her partner Ryan Genz on the runway after one of their successful showings. The team has worked together for over ten years, since they were researchers together at the Interaction Design Institute Ivrea.

Attention to detail is evident in every stage of the prototyping process. Finding the right people to work with wasn't easy, though. Rosella remembers: "Engineers wanted to make functional circuits but didn't understand that we wanted them to be cute. Ryan was so frustrated that he learned how to design circuits himself. We designed the circuits like they are textile designs." When they were first working on the prototype for The Hug Shirt™, "an engineer came to us with a circuit design. It was 3 inches square, very thick and really ugly. We tried to explain, 'This product is called The Hug Shirt™. It's supposed to be cute because it hugs you.' We wanted the circuit to be heart-shaped. The guy just wouldn't do it. This was the first circuit we designed ourselves. It's heart-shaped and red, and if you look at it in the light you can see that the copper traces are in the shape of a butterfly." Rosella and Genz now do all prototyping in-house.

Whether working on a piece for a performer or their Hug Shirt™, CuteCircuit's focus is on bringing people together to share the magic. Their vision is to make something "amazing that creates surprise for people. These days people don't get surprised very easily. We are used to different, complex things bombarding us every day. It's incredible to create that amazement that gives joy to people. My favorite part of the process is when we are in the studio and I put the last stitch into a garment. We put the dress on someone and then it lights up. It gets really emotional. You can sense that everyone is feeling something. That's really awesome."

Fine Art—Driven by Concept

The role of the artist is to create work that will challenge the viewer and make them think. Their quest is to create work that will provoke, question, or simply engage. It is to connect form with meaning. Many fine artists have turned to smart materials to create work that illustrates some of today's issues surrounding technology, communication, and privacy in a tangible way. Their work is often the first encounter that the general public has with an emerging technology beyond what they read about in the media. Installations created as displays of public art have used smart materials to create amazing interactive experiences that connect the viewer with the art, and therefore connect them with the materials, bridging the gap between research done in a lab and a consumer product.

Often conceptual in nature, these works may involve creating new tools and methods of making to build an installation, create a performance piece, or simply craft a sculpture. The key for an artist wanting to execute their vision is finding and mastering the materials that will yield the desired result. Their approach will be a combination of learning about a new material and expressing the meaning of their work. The fine art approach is driven by concept, but is manifested in materials.

For fine artists, as with many designers, the process starts with a seemingly random phase of experimentation that involves observing and experimenting with a collection of ideas, and working with different skills. Gradually certain ideas or problems will become the focus of the research and will generate a particular problem or thesis idea that is then refined into the driver toward a series of solutions. The artist will then begin a series of explorations. Each work will examine one aspect of the idea(s) while generating additional aspects to be explored in further works.

This cyclical way of working is one of the most important aspects of the creative process. No matter what discipline an artist or designer works in, this process of experimentation, evaluation, and evolution of ideas—learning from your mistakes while moving forward with new ideas—is the driving force behind a successful creative process.

The fine artist's approach is an exploration of concept through the use of materials to express and explore a series of solutions relating to a specific topic. These materials can range from the traditional ones to emerging technologies. The work of fine artists is a narrative; for the viewer it's between the art and their experience; for the creator it's from piece to piece. The two case studies that follow demonstrate how the exploration of materials can yield very different results.

Both artists have had long careers that have advanced and adapted as technology itself has evolved. Barbara Layne's work explores how technology and textiles interact, and is driven by her undying love of fabric. Maggie Orth brings up very interesting questions regarding sustainability and technology. Her choice to retire from electronic textiles will no doubt spark a greater awareness of this subject—as designers, we all have a duty to understand where the materials we use come from, how they are sourced, and what their impact is on our planet.

Barbara Layne

Studio subTela

Barbara Layne combines traditional materials and digital technologies in the interactive textile works she creates. A professor of fibers at Concordia University in Montreal, Canada, she is director of Studio subTela and participates in the Interactive Textiles and Wearable Computers research hub, Hexagram-Concordia Centre for Research-Creation in Media Arts and Technologies. Layne's research is supported by numerous foundations, including the Canada Council for the Arts, the Canadian Social Sciences and Humanities Research Council, Hexagram, and the Conseil des arts et des lettres du Québec. She lectures and exhibits her work internationally. Her long and decorated career spans the birth of e-textiles, and she continues to innovate and explore with new investigations.

"Carriers of culture"

Layne describes herself as someone who is interested in expressing herself through textiles. She thinks of textiles as "carriers of culture" and is intrigued by how they transmit information. From this context she poses the questions that have driven her work for many years: "What is next for textiles? How can I best represent the textiles of our era?" Her answer is, "Through digital technology." Her passion is experimenting with embedding technology into fabrics. She adds, "I'm also interested in the X and Y axis of textiles, the warp and the weft, and how they relate to a circuit board, which is often on the X/Y axis."

Her early work was all painstakingly hand-woven with conductive threads and LED components that she had to jury-rig by removing the hard-wire connectors and retrofit with conductive thread so that they could be incorporated into her fabrics. Once the fabric was constructed, she would then be able to cut and sew it into a garment. This entire process could take as long as six months for a single garment. Perfecting the weaving process with the conductive thread and LEDs took over five years, and countless trials. "It was so slow and so precise that we weren't moving fast enough. We wanted to speed it up a bit, so we started taking a few shortcuts and started using ready-made fabrics. We still rely on what we learned from the hand-woven work. For example, I'm pretty sure that we're the only people who have figured out how to weave a one-layer fabric using non-insulated conductive threads within a structure so that they don't short-circuit. We know how to make them not touch by manipulating them within the structure of the weave itself."

One of the unique features of Layne's work is her ability to source and use different types of equipment in the studio to work with the electronic textiles she creates. She talks extensively about how finding a new machine or component can change the direction of a piece, and even her research. Changing the wire pins on traditional LEDs to soft legs with conductive thread was extremely time-consuming, so when she

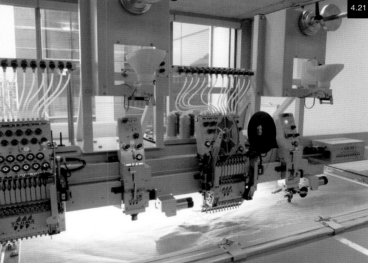

4.19
Designer Barbara Layne embroiders intricate patterns of electronics onto the surface of existing textiles to create unique e-textiles.

4.20
The Tajima laying machine—similar to an embroidery machine—is used to apply yarns or ribbons to the surface of an existing textile.

4.21
Here Barbara and her team experiment applying conductive materials with the laying machine.

4.22
Barbara's fabrics are hand-woven with conductive threads and LED components. At first she had to modify existing LED components so that she could weave them into her textiles; now she uses bead-shaped LEDs with tiny holes through which the conductive thread passes.

4.23
Once complete, the woven LED fabric can be cut and sewn into garments. This fabric is used in Barbara Layne's *Currente Calamo* collection, which used digital communication to display letters on each garment.

4.24
In addition to displaying letters, the light-up fabrics can also display numbers and symbols. When not in use, the LEDs look like glass beads or sequins on the surface of the fabric.

4.25
Tubular LED lights are sewn throught the weave of the white fabric as seen in this detail.

discovered a company that produced an LED with two tiny holes through which a conductive thread could pass, it made working with LEDs akin to beading or applying sequins. Most recently, she has purchased a Tajima laying machine, which provides an automated way of applying wires, conductive threads, or other materials to the surface of a fabric, eliminating the need to use glue to sandwich wires or flexible cables between layers of fabric. It uses a process called couching (sometimes called laid work), an embroidery technique in which yarn or other materials are laid across the surface of the ground fabric and fastened in place with small stitches of the same or a different yarn. "It's a huge table with the laying head device on top. Similar to an embroidery machine, it couches the conductive material in place on the surface of the host fabric." Layne sees this machine as "the newest thing," and an opportunity to experiment—she anticipates that the work will become very inventive.

Because she is not a fashion designer, her design process revolves around the concept of each piece. "The process starts with me thinking up an idea, generally a concept. I ask 'What are we going to do with this dress? Or this jacket? Then I source a pre-made pattern that will fit my concept. I am not a patternmaker, so I purchase a pattern then alter it to fit my concept. First I have to say that I work on a team. I've been working with an electrical engineer for over eight years, Hesam Khoshneviss. After we get the pattern,

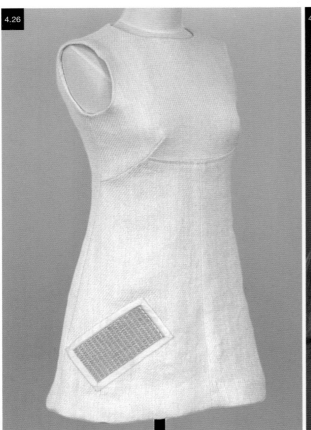

4.26
The White Touchpad Dress combines natural fiber woven with microcomputers and sensors to create surfaces that are receptive to external stimuli.

4.27
The *Currente Calamo* dresses are a system of hand-woven garments that include their own flexible LED message board.

4.28
Each dress has its own unique Bluetooth address. Transmitting to that address will change the images and texts on the LED array in real time. Tweeting the dress will reflect your message.

4.29
Barbara Layne works with electrical engineer, Hesam Khoshneviss, her collaborator for over eight years.

4 PEOPLE AND PROCESS

Hesam will start working on the technology. We always change the designs as we work; there's always a new way of doing things or a new level of technology involved. It's a lot of trial and error. He'll be busy in his area working out the circuits, buying the sensors, and sorting that out.

"I'll make the dress out of muslin, just to see where things might go, but we won't complete it with the electronics until we understand where the cabling will go and have made sure everything will work. We usually put the heavier electronics down by the hem. We'll test everything out in muslin and then make the dress. The electronics go in last."

Each one of their pieces will go through many iterations before the final piece is assembled in the fabric. Layne is currently working on assembling her work for a retrospective.

Considering what she could share from her experience with others who are just starting to work with these materials, Layne suggests two things: "Build a network of friends who are doing the same things you are; you will learn from one another. I was fortunate that way and have learned so much from my colleagues. The other thing is to absorb as much information as you can from the Internet. It's amazing how many resources there are out there, from DIY sites and lists of material resources." Layne also stresses the importance of learning how to operate all sorts of machines. In her studio there are digital printers, digital looms, digital embroidery machines, hand-weaving looms, the new laying machine, and standard

4.29

industrial sewing machines, which can overlock, cover stitch, straight stitch, zigzag, and so on. "The more knowledge you acquire, the easier it will be for you to be free when you are creating your designs."

For Layne, the inspiration for her work is simple. "The difference between me and other people is my love of cloth. I'm not so interested in the wearable aspects, it's more about what cloth can do and say, how it can change individuals and the environment."

Maggie Orth

artist and technologist

Maggie Orth is in transition. She is an artist, writer, and technologist who has created interactive and electronic artwork in her studio in Seattle, Washington, for over ten years. A pioneer in electronic textiles, her artworks include textiles that change color under computer control, interactive textiles, and robotic public art. She holds a PhD in Media Arts and Science from MIT's Media Lab, and a Master of Sciences from MIT's Center for Advanced Visual Studies after starting in the arts with a BFA in painting from Rhode Island School of Design.

Orth creates her artwork in the context of the company she started in 2002—International Fashion Machines, Inc. (IFM), where she has developed the creative, technical, and commercial aspects of electronic textiles. She has written numerous patents, conducted research, developed her own technology, and designed products. Her work has been exhibited in many shows and museums, written about in many publications, and has received some of the field's highest awards. But she has begun to question the future of electronic textiles and their value to society.

Orth's process is a melding of engineering and fine art. She has mastered the engineering skills necessary to be creative in a highly technical field that is constantly in flux but also resistant to change. She is both expressive about and frustrated by how difficult it is to achieve her visions. "My position on electronic textiles is not extremely positive. Electronic textiles, in my

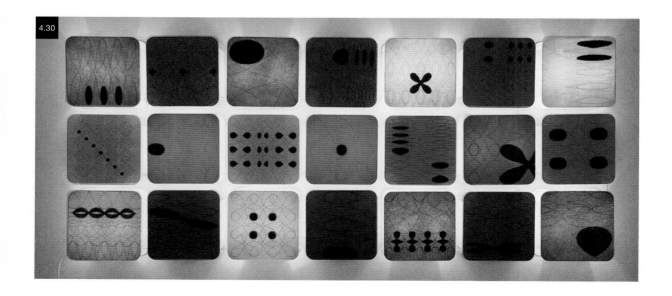

4.30

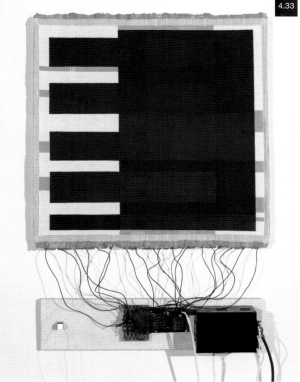

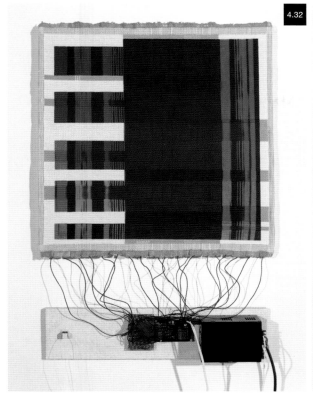

4.31

4.30
The *Petal Pusher* project is an interactive textile and light installation that uses a patented electronic textile sensor made up of yarn twisted from fine metal fibers that are charged with small amounts of electricity that react to the human body to complete a circuit that will brighten and dim the lights.

4.31
Maggie Orth in her studio. Her company International Fashion Machines, Inc. creates textile-based home products and one-of-a-kind color-changing artwork.

4.32
The color-changing textile pieces are created from conductive yarns woven with electrodes and colored with thermochromic inks that make a changing pattern that is controlled by custom-written software.

4.33
Blip, created in 2010, is one of seven pieces in Maggie's "moving toward stillness" series that explores the the magical properties of the electronic materials that bring life, energy, and motion to something that is normally still.

4.34

Two images of the piece *100 Electronic Art Years*, created in 2009, which explores the indefinite lifetime of electronic color-changing textiles. The combination of conductive yarns with thermochromic inks and woven electrodes, which are driven by electronics and expressive software. Seen here are two of the color combinations created when a current is introduced to the piece.

experience, are actually electronically limited materials. I have accepted over the years the electrical limitations of my materials to have a textile-like experience, to preserve the textile feel. There is simply a fundamental limitation to what can be done with these things. You can make sensors, you can pass some data around, you can light an LED, and of course you can have an infinite number of small iterations and variations on that. My work has really been about exploring the possibilities of these materials given these limitations."

Programming and materiality

When Orth was first starting out with her color-changing textile works, she "wanted to explore the aesthetic possibilities between programming, computation, and materiality. My question was, 'Is there something meaningful between how the piece is woven and how I program it? Is there a place where the software becomes related to the materials?'" Her body of work with color-changing textiles has explored this central issue for many years. "I was really interested in a textile that could hold thousands of patterns and could change over time. My idea was that once you start programming something, it can do anything." But as her work progressed, she began to embrace Michelangelo's truism, "Every block of stone has a statue inside it." "When you make a programmable material system, you have all these ideas of what you are going to do with it beforehand. But there

is a continuum malleability in programmable material systems—on one side you have totally neutral systems, like an RGB display where you can program it to look like anything. On the other side, you have something like my textiles, which are more toward the material side of things, where there is a limited amount I can control with the programmability."

When it comes to insights into her work process, Orth explains, "The materials aesthetics of my pieces have always been a starting point. But I collaborated on everything. It was all collaboration. In fact one of the challenges of my work has been the enormous amount of collaboration that was necessary to complete it. I had to hire engineers and programmers and other technicians. And the most challenging part about working on collaborations is that you have to work with other people. It takes a tremendous amount of time to get everyone working together on a project, and communicating well." However, Orth adds, "The great thing about collaboration is that you can end up with something that far exceeds your scope as an individual." She compares it to filmmaking, where projects are so huge that it is almost impossible for one person to have complete creative control over a project. If you were to control everything the payoff could be huge, but more than likely your work will be compromised because it would be impossible to master every aspect of the work involved in today's complex media projects.

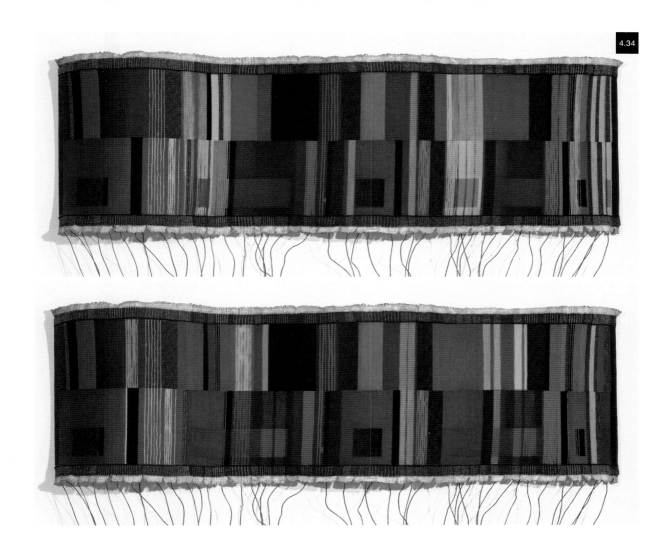

Now that she has concluded her work with color-changing textiles, Orth has begun to explore ideas related to the environment—in particular, the environmental impact of e-textiles. Having carried out research into the topic she had come to the conclusion that, "I didn't feel that making electronic textiles was very good for the earth." Her solution: She started writing.

Part of Orth's art process has always been storytelling, so it was a natural transition for her to take up writing fiction to give a voice to her thoughts on how economics, design, and technology all influence our environment. She describes her excitement about this new direction:

"I feel really free of the physical burden of my work, which was; 'Where do I keep it?' and 'Where can I find that part?' The scope of the physical burden in my life, of my work, was just extreme." No longer will she have to spend six weeks sourcing a square white LED button to refurbish a piece for a show, a task that looks in hindsight to be both important and irrelevant. Once freed of the day-to-day chores of making physical pieces, her creativity has rebounded. Her fiction focuses on people, nature, and technology. And she is currently working on her first novel, which is about a boy who goes on a spiritual quest and encounters an alligator from a computer game.

Fashion—Driven by Aesthetic

Fashion is the pulse of our society. Constantly changing, it adapts and reacts to how we are feeling, and reflects what we are thinking. The ebb and flow of fashion is in constant flux. It is the only form of design that relies on a wearer to interpret and interact with it in order to bring it to life. The designer has only so much control over the work, before it is relinquished to become part of a larger collection in an individual's wardrobe.

Fashion designers are both artists and visionaries, constantly creating looks that capture a moment in time or make a statement about a future yet to be realized. Fashion is both fantasy and practicality, and perhaps for this reason it has a larger following than any other form of art and design. It is living art.

A fashion designer is constantly absorbing everything in their environment. Taking it all in, they process this information and create an aesthetic from which a series of looks emerge. The designer's vision combines intellectual and visual research along with ideations to create a narrative, and successful designers find a unique voice with which to tell their story. The goal of this narrative is to trigger an emotional response—to make you think, feel, or question. Every step of the fashion designer's process is *driven by aesthetic*.

The fashion design approach is both visionary and reactionary. By nature it is full of contradiction. The designer's process is as much about a personal quest for a unique voice in a saturated media landscape as it is an exploration of conceptual ideas that question the meaning of clothing.

Materiality is fundamental to fashion. While scholars engage in the study of fashion in terms of how fashion is materialized—its persistence, endurance, transformations, and its relationships— designers focus on its sensory effects—tactility and aesthetics. The work of the designers in the following two case studies focuses on a shared exploration of experimental fashion and performance outside of commercial parameters. It prompts discussion and engages our intellect. Today's climate of collaboration and interdisciplinary thinking has opened doors for fashion to emerge as a leader in a contemporary practice that embraces technology and digital media.

Ying Gao

conceptual fashion designer

Ying Gao, an accomplished conceptual fashion designer and professor, draws a line between art and design, but confesses that this line is starting to become blurry. "Designers work on an object, something concrete—well, maybe not concrete, but a tangible object. In my case it's garments. Being an artist is totally something else. Maybe I'm being a purist but when we're talking about working on garments, I'm not creating art, unfortunately for me." Gao's work is nevertheless very artful and provocative.

She studied fashion design at the Geneva University of Art and Design and then moved to Montreal to pursue a master's degree in Communications and Multimedia from the University of Quebec, remaining in Canada for the next 20 years. Now she is back in Switzerland, having just accepted a post as Head of Fashion at her alma mater. Gao reflects on the time she spent studying multimedia in Montreal as "a period of my life where I learned a lot of technical things. I also learned from a theoretical point of view a lot of things about media that helped me build up another kind of thinking. I didn't work in fashion for two years."

Experimentation

The combination of these two disparate degrees and their approaches to design have formed her creative approach. When asked about the evolution of her process and what she took away from the combination of the two, she replies, "I think the most important thing is experimentation. Because of the reality of the fashion market and the fashion industry, fashion designers are not taught, or have not been asked, to be brave enough. They have become very practical. We have lost this critical thinking that people have in other fields such as media design and cinema."

Gao often finds the spark for an idea in books or movies. She has always been interested by her environment, the things in her immediate surroundings, and the somewhat intangible elements of these—light, sound, and air. When a movie or a novel captures something that speaks to this interest, the concept for one of her collections is born. In 2004, Gao was reading a book about Archigram, a London-based avant-

4.35

4.36

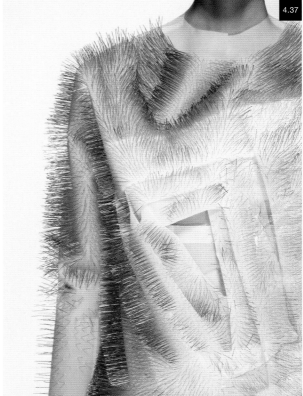

4.37

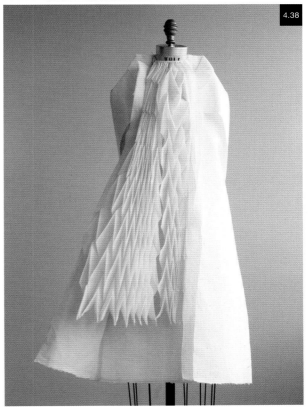

4.38

4.35

Ying Gao, a fashion designer and professor, in her studio. Her work questions assumptions about clothing by combining robotics and fine fabrics to create interactive pieces that transform their physical contours.

4.36

Incertitudes consists of two interactive garments covered in dressmaker pins whose movements are controlled by electronic devices that respond to a spectator's voice.

4.37

In this detail of the piece, the sharp ends of the dressmaker's pins are evident. The *Incertitudes* garments both engage the spectator and create a level of uncertainty.

4.38

In *Walking City* (2006), the fluid movement of breathing is simulated by sensors and a pneumatic mechanism sewn into the cotton and nylon fabric.

garde architectural group formed in the 1960s that drew inspiration from technology in order to create a new reality that was solely expressed through hypothetical projects. One of their projects, *Walking City*, which inspired Gao's collection of the same name, was a collection of intelligent buildings that were giant living pods roaming cities. Gao explains the synergy between their project and her inspiration: "They were working on some utopia projects such as inflatable cities and walking cities. They were pointing out some of the failures of our society, and demonstrating that we were more and more into consumerism. It inspired me to work on a utopia project, and a possible idea was some inflatable garments. But I didn't want them to look like the Michelin Man. So I worked on pleats. I studied a lot of origami. And I combined both into an inflatable system."

To create her work, Gao has worked with a robotics designer, Simon Laroche, for over ten years. Together they create the interactive clothing that Gao conceives. "I bring up the idea and Simon and I work on the technical side together. I do have assistants, and because I teach at a university, I believe it is very important to have my students involved with these kinds of projects. The problem is they can't stay here long because they have their own projects and eventually they graduate. It's never-ending—spotting new students, training them, the coming-and-leaving process. I could hire people who could stay and have a more stable situation, but as a professor,

I think it is really important to have the students involved."

Once she has the spark of an idea, she starts sketching, using this time to work out exactly how she wants her collection to function. "The most recent example is the *No(where) Now(here)* project, which was inspired by an essay. It's from *Esthétique de la disparition* (The Aesthetics of Disappearance) by Paul Virilio, a French philosopher. He talked in this book about the specifics of disappearance, a lot about being gay, the moving image and the notion of absence, including speech and light and other things that have interested me for a long time. From that, I decided to work on a pair of dresses that appear and disappear according to the eyesight of the spectator. This was my first idea. It was very vague, and it really seemed like nothing. Then I sat down and started sketching. I needed to decide not only what kind of garment it would be but I also needed to think about the interactivity. Then when the whole thing was almost done, in my head, I talked to Simon about the idea."

Gao is amazed that Laroche is able to solve so much with technology—he always seems to find a system that makes her projects work. The *(No)where Now(here)* project took the pair two years of part-time work to complete.

When describing her sketching process, Gao explains that she sketches both in two dimensions in her sketchbook and in three dimensions using draping and patternmaking with the actual fabric—

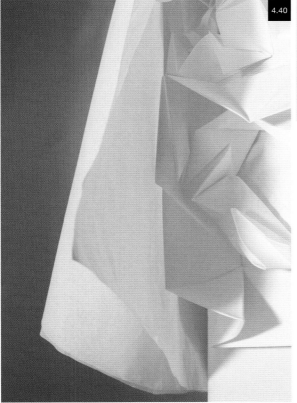

4.39 & 4.40
Walking City (details). The piece is composed of inflatable sections that create a playful breath-like motion in the intricately pleated fabric. It is a homage to the British architecture collective Archigram which imagined mobile and inflatable habitable structures in the 1960s.

4.41
The two dresses that form the *(No) Where (Now)Here* project are created from photo-luminescent threads. The dress has embedded eye-tracking technology and robotics that create movements in the surface of the dress; these react to the spectator's eye movements.

Super Organza, which she works with for all of her projects. She works the silhouette of each dress in detail until she has it the way she likes, then adds the electronics. She never buys fabric with the wiring and electronics already integrated. Since each of her projects is so unique, the technology needs to be built from scratch for each one. She explains, "Sometimes we'll have to 3D print something to get it prototyped. For example, we didn't find the appropriate motors for the last project, so we had to print our own motors."

Prototyping

The prototyping phase is the longest part of the process, and Gao describes finding the correct wire to make the jellyfish for one of the *(No)where Now(here)* dresses as one of the more difficult obstacles she has had to overcome. "I wanted to reproduce these specific movements with my dresses. It was really difficult. I can't count how many prototypes we made in order to eventually obtain this. We tested 20 different kinds of wire. Each jellyfish is made of a very very thin wire structure, and the thickness of the wire is very important because it affects the movement. Every time we tested a wire, we had to make five jellyfish so that we could put on the motor and see the movement. Within 30 seconds we would know if it worked or not. Just 30 seconds and then it's like, 'Nope. We've lost too much time. Let's restart.' It was just trying and trying."

Gao confesses to being "a bit stubborn" about this sort of thing. "I don't want to make gadgets and I don't want to compromise. For me, it's very important to make things perfectly. I'm not saying my garments are perfect; I'm saying that I have a goal to reach and failing to reach it is not tenable." Reaching her goals without compromise is paramount. Her work is truly beautiful, so her hard work and perseverance is paying off.

Finally, Gao shares a bit of advice for designers just starting out. "Be brave enough in life to be curious in order to experiment with things. We all have to pay a lot of attention to knowledge. It's easy to have crazy ideas, but you need to have technical knowledge to make things happen. If you don't know how to make things happen, then the crazy ideas are useless. It is important for the new generation of students not to neglect the technical aspects of the world. I know this sounds very practical, but we have to be grounded, we have to be realistic in this profession. You need the knowledge to be able to experiment."

4.42

(No)Where (Now)here dresses explore ideas of presence and disappearance. They create an experience of alternating clarity and obscurity for the viewer, whose gaze alters the surface of the garments.

4 PEOPLE AND PROCESS

Lynsey Calder

CodedChromics

CodedChromics is an exciting project headed by Lynsey Calder that explores color-changing dance costumes. Calder's background is in textile design, but today she is an investigator and researcher at Heriot-Watt University in Edinburgh, leading a team that is exploring the interface and interaction between smart textiles and smart or pervasive environments—an advanced computing concept where computing is made to appear everywhere and anywhere.

Her current project is "Smart Costumes: Smart Textiles and Wearable Technology in Pervasive Computing Environments." I caught up with Calder in her studio to discuss her creative process, collaboration, and her latest project, the color-changing tutu.

When she first started working on the color-changing tutu project, there were a lot of discussions around the reasons for using costumes in performance, and what role the

4.43

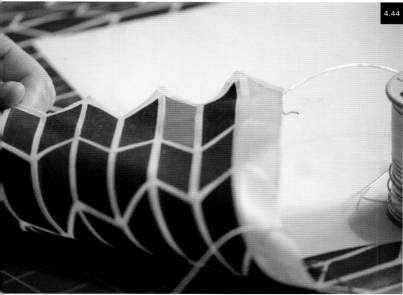

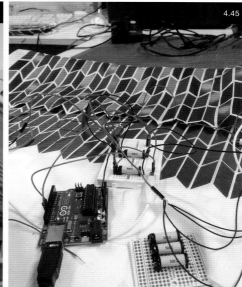

4.43

Lynsey Calder's concept board with butterfly wing studies used in the inspiration of her color-changing tutu costume.

4.44

Detail of the fabric's geometric pattern printed in color-changing inks. With the introduction of an electrical current, the inks change in hue from purple and blues to magenta and pinks.

4.45

Once wired, the pattern of the printed fabrics is controlled by an Arduino Mini programmed so that the fabric responds in time to the music used in the dancer's performance.

4.46 & 4.47

Lynsey painstakingly hand-soldered each of the wire leads onto the fabric. Conductive paint is used to create the connection point and completes the circuit.

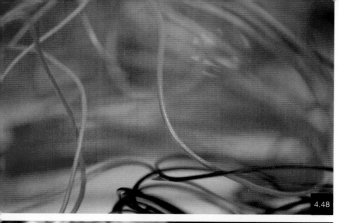

4.48

4.49

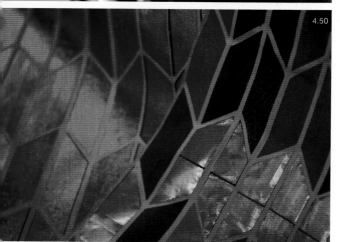

4.50

4.51

costume plays in the overall performance of a dance. "The concept that I am developing explores the evolving textile topography of a dance performance through color-changing costumes, affecting both the audience and the dancers. Color can be very evocative and the color-changing element of the costumes adds another layer of complexity to the choreography of the dance."

Beauty and technology

Calder sees her work as an exploration of beauty and technology, and how the intersection of the two come together to create new and exciting experiences.

"I love the idea that technology and science can be beautiful and visually engaging," Calder says. "My inspiration has always been to make something beautiful from technology. But on this question I can see both sides: On the one hand, I can see the need to stay pure and to have a part of our lives that remains unplugged; but the attraction and fascination of new technologies will no doubt be something that takes over more and more of our lives."

Working at a university research facility is a different experience from working in your own studio. The academic environment is conducive to building collaborative and cross-disciplinary teams of researchers, and the tutu is no exception. Calder relies heavily on a collaborative team of academics and artist residents at the university,

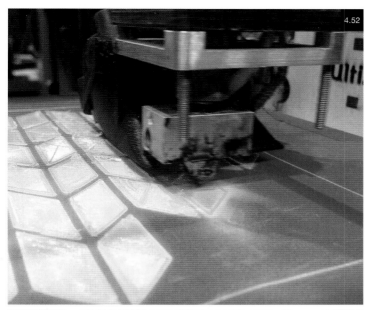

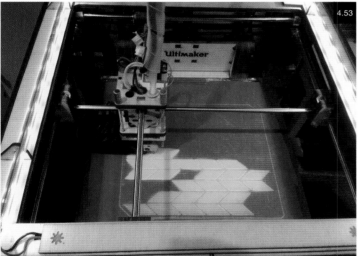

4.48 & 4.49
Detail images of electrical wires connecting to printed fabrics.

4.50
Power running through the surface of the fabric creates a vibrant, color-shifting appearance.

4.51
The geometric printed pattern designed to change colors but not short-circuit.

4.52 & 4.53
Detail of printing machine applying conductive ink to the surface of the tutu's fabric.

4.54
The final prototype on display. The Arduino controls and wiring are completely visible, illuminating the creative process.

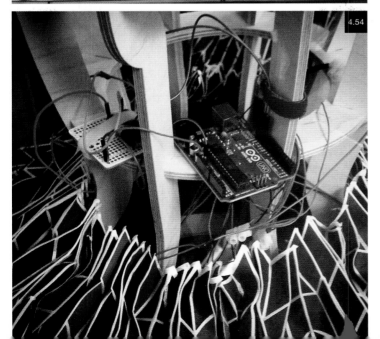

149

where she is a lead investigator, and at others across the UK. Once she has decided on the scope of her work, she relies on a team of people to help her investigate solutions.

"I'm on a team with Professor Ruth Aylett and Dr. Sandy Louchart, who are behind the pervasive computing side of things. However, the main hands-on help has come from two other practitioners on this project—Dr. Sara Robertson, who is a lecturer in craft innovation and smart materials, and Dougie Kinnear, who is a jeweler currently working as a researcher at Dundee College of Art and Design, both of whom are crossing the design–technology interface," explains Calder. Kinnear helps with building the circuits and programming, while Robertson, whose research on designing with liquid crystals (her PhD thesis is 'An Investigation of the Design Potential of Thermochromic Textiles used with Electronic Heat-Profiling Circuitry') has underpinned the building of many of the components of the tutu, including the heat sinks, which allow the electrical current to dissipate heat into the surrounding area. Calder's role is chief builder and she runs the project on a daily basis.

"I have used the skills that have been demonstrated to me by others and continue to build on this knowledge. I am also an advocate of 'open-source design' and find that the communities of designers/makers online have helped in the development of my ideas and concepts. My main role has been to produce interesting and exciting working prototypes of color-changing fabrics that respond to external and integrated elements, and then develop these ideas into a finished costume, which is now under construction."

Thermochromic inks change color when exposed to specific, pre-programmed temperature changes. The tutu is a combination of screen printing with a mix of three different temperature-threshold thermochromic pigments. The fabric they developed is printed on the face of the fabric with these inks and has a foiling on the reverse. On top of the foil are heat sinks made from copper foil tape and surface-mounted electronics, which are wired together in rows.

Prototyping

The first prototype took almost a year to construct. Just creating the fabric took many trials. Now that they have the fabric, the team is creating a full-scale working prototype. The next step is to test it on a dancer for comfort, agility, weight, and heat generation. When talking about the prototype, Calder explains, "One of the obstacles we have to overcome is the power source. Currently it runs on a large number of 9V batteries, so either we have to look at adapting it to another battery type, or redesign the costume to accommodate the batteries effectively." One iteration of the costume leads to the next, each building on the successes and failures of the previous model.

Inspired by an origami-folding pattern, each

trapezoidal shape is programmed to change color according to the increasing or decreasing heat provided by the electrical current. To add another layer of interest, a fluorescent pigment is mixed into the thermochromic pigment. When mixed together, these two dyes create a gradient glow effect. Not only can the fluorescent inks be activated by UV/black light on stage but, as the thermochromic inks are heated, the fluorescent colors also begin to appear. This gives the costume the ability to be emissive and non-emissive, depending on the lighting, making many lighting effects possible. Arduino and a simple setup of relay switches control the costume's phase-changing element.

The academic environment supports the free flow of ideas with funding and space for work like this. That said, Calder can see this research may lead to applications that involve safety and alert systems, especially ones that involve children. "Color is a universal language, and could be used as a code in a school uniform, or on backpacks. There is potential for signaling devices for learning in situations where individuals find conventional language challenging. I can also see that research in this area could be applied to the medical industry, perhaps as a signaling trigger or on a completely different scale for internal surgery."

Although influenced by her love of science fiction, particularly the writing of Iain M. Banks, Calder has found that her time exploring ideas in the research lab can fuel real-world applications.

4.55
The finished prototype of the color-changing tutu undergoing color transformation.

4.56 & 4.57
Color variations in the tutu, showing different degrees of electrical current flowing through the fabric and how it affects the color range of the printed pattern.

Architecture—Driven by Site or Solutions

Architects approach a project with a very specific set of parameters. They approach each problem from the perimeter, surveying the issue as they would survey a site and create a plan. More conceptual and experimental architecture tends to be removed from the user experience, and focuses more on the solution, looking for innovative ways of applying theory and upending previous thought processes to create newness. Creating a solution, the conceptual architect then figures out how it works. The architectural approach is therefore solution driven.

This type of architectural process tends to divorce itself from the needs of the user as a primary driver of the design process. Whether the problem is posed by the designer or a client, the architectural approach is very conceptual in its initial stages. Concentrating on form and space, abstraction and experimentation are encouraged, with a free flow of creative ideas. Once conceived, it then goes through a stage of model making and prototyping, when the design is rationalized in order to adapt it for the user and to make it more buildable. This approach to design is different than other disciplines, and separates the initial creative process from the execution phase.

The following case studies are extremely different in both scale and scope, but their commonality is their approach. Looking at the result of their projects, you see that these designers approach their work with an intellectual exploration of an idea, and then rationalize that idea through model making and materials.

Summary

Whether making models or building full-scale installations, the two design firms discussed on the following pages illustrate how the architectural design process is divided into a conceptual exploration, followed by a rationalization of the design to execute a functional solution.

The initial ideation phase explores solutions in an open and free creative process that focuses on form, abstraction, and materials, and often uses computational modeling as a sketching tool. In later stages, the design is rationalized, evolved, and retooled so that it can be executed. The focus of this approach is always to achieve a solution, whether to a practical problem, or to a conceptual question. Materials are explored both during the ideation and execution stages, and can inform design decisions at any stage.

Some of the most inventive solutions and materials applications happen when approaching design through the architectural process.

Daan Roosegaarde

Studio Rosegaarde

Studio Roosegaarde is a dream factory for interactive design. Its founder, Daan Roosegaarde, shies away from defining his work as art or design. Speaking at TEDxBinnenhof in 2012, Roosegaarde explained, "I am not a designer, I'm a reformer. I am fascinated by the idea of merging imagination and innovation into one." He calls what he does "personalizing and customizing the world," making the world more understandable, interactive, and open.

When speaking about his work, Roosegaarde explains, "We use technology to bring things to life, to open things up for other people. It's about the interaction between people and objects but

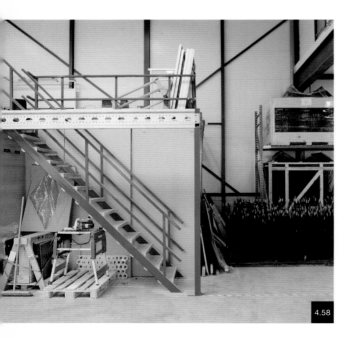

4.58

also between people amongst each other. So it's not about high-tech but more about high-social." Most of the work created by Studio Roosegaarde, when not on display in museums and fairs, consists of installations in public spaces, like pedestrian walkways or highways.

When talking about the role of art and design, Roosegaarde shares a saying they use in the studio, 'It's not about the color of your lipstick, it is about whether you can kiss or not.' In other words, it's not about the object, it's about the behavior that it generates. This is pragmatic. He continues, "The role of the designer is to create new niches, to create links between the old failing world and the new world." He believes that once an idea is introduced, the market will follow. This is the driver behind many of the studio's conceptual installations.

The artists and the technicians

Studio Roosegaarde consists of two groups, the artists and the technicians. The artists make the "artist's impressions"—another way to describe ideations—and source and manipulate the smart materials used in their projects, while the technicians develop all the studio's own technology. Everything made in Studio Roosegaarde runs on its own technology.

"It is very important, like Rembrandt had his paint, that we have our own microchips. It is our way of expressing ideas. Developing your own knowledge and technologies gives you total

4.60

4.61

4.58
Studio Roosegaarde: the *DUNE*
project is seen before installation
on site in the lower
right of the image.

4.59
In the studio designers work
on a variety of projects from
interactive spaces to architecture
and fashion. His *Smart Highway*
project uses glowing paint
that changes color with the
temperature and lighting that
charges during the day and
shines at night.

4.60
Roosegaarde Daan travels
constantly and is always in
search of inspiration for his
many design projects. Boxes of
research materials from his many
trips line the office walls.

4.61
A self-described "hippie with a
business plan," Roosegaarde
Daan works on the floor of his
studio. With locations in both
Shanghai and Waddinxveen, the
Netherlands, he is constantly
in motion, striving to realize his
many visions.

freedom. You're not tied to a standard LED or sensor," Roosegaarde says. With this approach, the work is not hindered by the constraint of existing components. He goes on to add, "It's all about the idea in my head, the flavor in my mouth and looking for the ingredients you need to create that flavor."

Invisible technology

The team knows that it is also very important that everything that is generated works perfectly so that the technology becomes invisible. The work needs to be about the question, about what it stirs up; it needs to transcend the medium. "My work has aesthetic value because that way I can lure people in; the users have to be unconsciously seduced. Once they're inside, I can enter into a dialogue with them, manipulate them, make a point, make them conscious of something, make them do things they would never otherwise do, but that they experience as natural," Roosegaarde explains. "Technology is so important now because it's such an important part of who we are, because it has so many mediating qualities, and I also find it exciting because you can make things that are never finished, but that encourage interaction."

The "Intimacy" project is an example of luring people into a discussion about technology as our second skin through the use of materials. Until starting this project, the team was unfamiliar with fashion. The dresses they have created use a film that changes from white to transparent based on how excited you are. The faster your heart rate, the more transparent the dress. But the dress is only the manifestation of the issue which questions, "How can technology become more tactile and more intuitive?"

Roosegaarde was provoked by this thought: "Technology is our second skin, our second language, the way we communicate our experiences and our information. So why are we alway looking down at these iPhone screens the whole day?" Intimacy started with the idea of hiding and showing. "I always talk about sensual technologies, extensions of your skin, and I thought, now let's use it in a really extreme way: let's use it in fashion. Fashion, after all, is very much about the tension between hiding and showing. I think we're developing toward a world in which local intelligences will 'talk' to each other through wireless connections; they could be objects or your cell phone. In this world of information, in which everything talks to everything else, there is a great transparency. It's really becoming an issue. But sometimes you want to hide things, and only show them later. What would happen if we used the concept of material becoming transparent and then opaque again, of showing and hiding, in fashion?" asks Roosegaarde.

The idea for the dress developed over time, but the inspiration came from a material. "One day I was in a lab where they develop flexible screens.

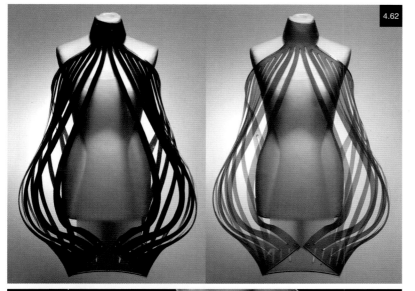

4.62

4.63

4.64

4.62

There are different versions of Studio Roosegaarde's *Intimacy* project. One dress is connected to the wearer's heartbeat, so the faster the heart beats the more transparent the dress becomes. Another is a man's suit that becomes transparent when the wearer is lying.

4.63

Detail of model wearing an *Intimacy* dress. The high-tech garment is made from a combination of leather and e-foils and changes from opaque to transparent in reaction to personal interactions.

4.64

Strategically placed blocks of e-foil reveal or conceal the model's body. The more aroused she is, the sexier the dress becomes. The garment creates a sensual play out of the act of disclosure.

I saw something lying in a corner. I asked, 'What's that?' 'Nothing really,' they answered. 'All it does is turn from white to transparent.' It was material that was 1mm thick. I was immediately enthusiastic about it. They let me take it with me. Then we continued to develop it with the manufacturer to make it more flexible and UV proof. Now we can dim it, too—at first it could only be on or off, transparent or white."

Public and personal intimacy

The concept behind the dress sparks a conversation about personal and private space. You always feel it when someone gets too close, this dress brings that conversation to the forefront. It is about making the energy of the body palpable and about the difference between public and personal intimacy.

When speaking about his creative process and what it has taken for him to get to where he is today, Roosegaarde says, "There have always been people who have said that what I want is impossible. It's my role to show that it is possible and to make it tangible."

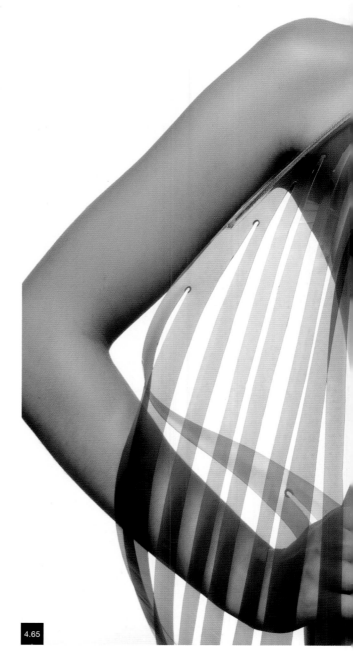

4.65

The Intimacy Dress is created from a combination of smart foils, wireless technologies, electronics, LEDs, copper, and other media.

4.65

Synthesis Design + Architecture and Jason Gillette

Fabric Images

In 2013, Synthesis Design + Architecture (SDA) from Los Angeles won the design competition for the Volvo "Pure Tension Pavilion." Their design was an easily deployable, freestanding membrane structure and portable charging station commissioned by Volvo Car Italia to showcase the new Volvo V60 plug-in hybrid electric car. To realize the Pavilion and prepare it for its official launch in September 2013 in Rome, Italy, SDA teamed up with the structural engineering firm, BuroHappold Los Angeles, and a fabricator, Fabric Images, Inc., an award-winning manufacturer of tension-fabric architectural displays and products. Together, the teams worked hand in hand on the creation of this experimental structure, using advanced CAD methods that included associative modeling, dynamic mesh relaxation, geometric rationalization, paneling, and material performance.

Alvin Huang, SDA founder, worked with a team of architects at his firm to conceptualize the Pavilion, which raised questions that the team found challenging. The structure had to be lightweight, easily deployable, and affordable; it needed to provide an iconic structure that would showcase the car, and it had to be able to travel with it. Their solution is a structure that can be assembled and disassembled in less than an hour, and can fit in the trunk of the car.

4.66

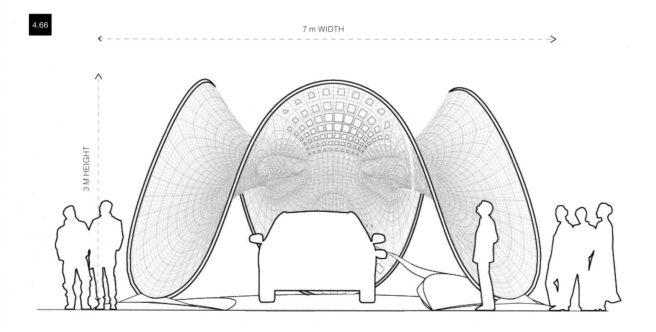

7 m WIDTH

3 M HEIGHT

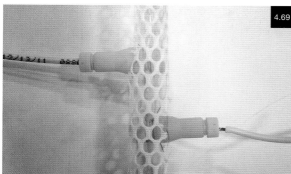

4.66

Rendering of the finished Pure Tension Pavilion showing the scale and form of the design in relation to the car and user. The project explores the possibility of carrying a solar charging station for an electric car in its trunk.

4.67

Each of the 100 plus pieces that create the final form for the Pavilion is laser-cut to precise dimensions.

4.68

A technician in the Fabric Images workshop lays out each piece of mesh that is then embedded with solar-charging cells and sewn together to create the pavilion's structure.

4.69

Solar energy collected in each photovoltaic panel is transferred through the electrical wires which are embedded in mesh channels that form the ribs of the pavilion.

4.70, 4.71, 4.72 & 4.73

Mesh panels are connected with precise detail and then wired (4.73) to collect solar energy that will in turn charge the car's fuel cell. A full charge takes 8–14 hours.

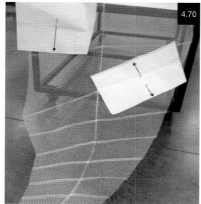

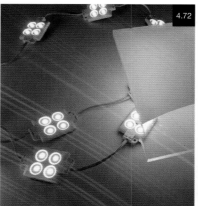

The team focused on tensioned fabric structures from the start. Fabric is lightweight, affordable, and easily collapsible. Once they had settled on a form that they felt enhanced the car, Huang explains, "We took the Pavilion a step further and began to think of it as a charging station for the car." It was the combination of these two ideas that helped secure them the winning entry.

The team then went through a process of rationalizing their design in order to build it. Jason Gillette, lead designer at Fabric Images, was in charge of executing the prototype of SDA's winning design. Gillette explains that what came to them was basically an incredibly beautiful computer-generated napkin sketch. "They used Rhino and Grasshopper applications to come up with this really complex, undulating form of tunnels creating other tunnels that create funnels. It was a really beautiful rendering, and they had essentially three months to figure out how to build it."

The Fabric Images team, lead by Gillette, started by testing fabrics, knowing that they needed to use a mesh to keep it as lightweight as

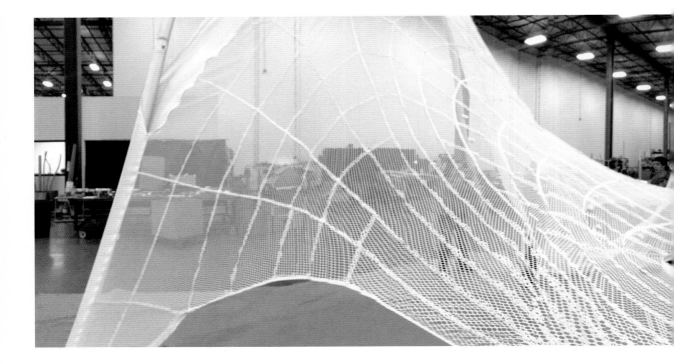

4.74

Panoramic image of the first prototype being assembled at the Fabric Images workshop.

possible, and to allow for use in windy locations. They tested three different types of mesh using dynamic mesh relaxation, a real-time form-finding process that utilizes computation to simulate force in materials to discover an equilibrium between form and force. "It's a prestressed cambering, which essentially creates a form by stretching a fabric across a frame that in turn will be pulled into place by the fabric to create the desired shape and geometries," Gillette explains. They settled on a mesh that was 70 percent air. Fabric Images created a three-part sandwich for the Pavilion's

skin, the mesh, then photovoltaic (PV) leaves made from a relatively accessible vinyl material that they could then attach PV solar cells to. In all the Pavilion has nearly 600 PV panels attached to its surface.

SDA supplied Fabric Images with a computer rendering of the 3D surface, which they were then able to use as a starting point to create a pattern from. They divided the surface of the structure up into manageable-sized shapes that could be CNC (computer numerical control) cut, a cutting process that is computer controlled for accuracy.

4.74

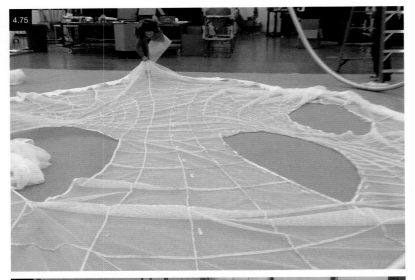

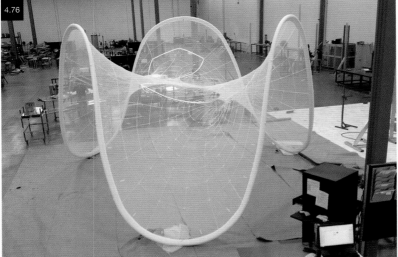

4.75
Mesh panels of the Pure Tension Pavilion constructed and ready for application to the tubular frame.

4.76
The first prototype is constructed with its mesh and frame structure united.

4.77
The first row of solar panels is applied to the structure and the designers assess the final prototype with the Volvo car for which it is designed. The scale of the structure is designed to complement the shape of the car.

4.78
The finished Volvo Pure Tension Pavilion photographed with the car. When fully assembled, it is designed to charge the car's fuel cell within 8–14 hours, weather permitting, and will collapse to fit into the car's trunk.

4 PEOPLE AND PROCESS

Due to the tight schedule, the team decided to prototype in full scale. Their thinking was, if they got it right, then they would be done and could meet the deadline. They broke the surface down into quadrilaterals and created a pattern that netted somewhere between 500 and 600 panel pieces that needed to be sewn together like a quilt. Hoping their calculations were correct, they constructed the first model and their pattern was accurate to within 1⁄8in (0.3cm) in the height and 3in (7.6cm) in the width. At that point it was just a matter of creating the nuance changes to achieve the desired form.

Creating the battery box

The wires for the PV cells run down the spines of the Pavilion and terminate in a battery box. The designers estimate that it takes somewhere between 8 and 14 hours to fully charge the battery. Gillette explains the process used to build it: "We created this battery box, essentially an aluminum-framed box that houses these big batteries and solar-charging hardware. We linked all the solar panels together on the actual surface and then channeled the electricity down to the fabric conduit that we custom-made into the base [of the Pavilion] and then into the battery box. The battery box is going to fill up with juice that will dissipate when you plug it into the car."

Gillette also explains how rewarding it was to work in collaboration with SDA to realize the creation of the Pavilion: "It was a fun process

4.78

because we were working hands-on with the architect, throwing sketches back and forth and coming to a final design that we were both happy with." The thing he learned most from the project was the importance of giving yourself the time and resources to experiment with materials, especially ones that you are contemplating using. "Really test them. It's almost like sleeping with the materials so you know them exceptionally well. Once you have a relationship with a material you can apply that same logic to new materials and then you start making better decisions and can start really exploring the potential of textiles."

Engineering—Driven by Scientific Exploration

Engineers take a practical approach to idea generation. Scientific exploration is often analytical and logical, with specific steps to follow in a managed sequence, with each step recorded and data analyzed.

Specific fields of design lean heavily on an engineering approach in order to be able to quantify the results of their innovations. A methodical, research-based design process has generated some life-saving innovations in the fields of medicine, safety and protection manufacturing, and space exploration. Engineers use a scientific approach when designing.

The engineering approach is founded on research. However, many art and design practices have research as part of their process, so what is the difference between an engineering and a design process? I consider that science, engineering, design, and fine art all exist on a continuum that travel in degrees from the concrete to the abstract. Each discipline utilizes research but in different ways; generating a body of research, though, usually falls toward the science and engineering side of the spectrum.

"What you learn from an experiment or an engineering process should hold true in many applications and be relevant to many stakeholders," says Dr. Lucy Dunne, a professor of fashion design at the University of Minnesota, whose own research and classroom instruction focuses on using an engineering-style approach. We have had conversations about the difference between design and engineering approaches, and agree that design usually seeks to solve a specific problem in context, such that the solution often does not apply anywhere outside of that context.

But there are gray areas where engineering and design merge. An engineering approach relies on evidence and quantified calculations to achieve explicit results, or to demonstrate improvements. At times, design may require this type of information—for example, when designing athletic apparel or footwear for competitive athletes. But not always. Often design is driven by other metrics, ones that are not quantifiable.

Dunne adds, "Engineering sees design as a part of the engineering process but I don't think design would necessarily agree. The separation is probably clearest in fields like architecture, where one person is responsible for the concept or idea [the architect], and another person is responsible for making it happen [the engineer]. This is also probably true in fashion, now that there is an increasingly clear distinction between creative and technical design roles."

The engineering approach is ultimately set apart from that of design by the process of identifying a specific question and answering it with a defined method of testing.

Sabine Seymour

Moondial

Often described as a visionary, Sabine Seymour plays many roles: designer, researcher, author, curator, and trendspotter. Her books *Fashionable Technology: The Intersection of Design, Fashion, Science, and Technology* (2008) and *Functional Aesthetics: Visions in Fashionable Technology* (2010) are seen as must-reads by anyone in the field of wearable technology and smart textiles. She is Director of the Fashionable Technology Lab and an Assistant Professor at Parsons The New School for Design in New York, and has curated, exhibited, and spoken about fashion and technology worldwide. She is also the Chief Creative Officer of Moondial, where her work focuses on intelligent clothing, concepts and creative direction for the integration of wireless technologies into clothing and equipment, wearable products, and trend scouting.

Seymour's ideas on the design process are very interesting, especially those concerning working with new technologies. "You need to create for yourself a huge knowledge-base of materials technologies, of processes, and of potential collaborators." She also speaks about working on projects where a particular material or technology doesn't exist at the onset of the process but becomes available over the course of development. Previous case studies have also discussed situations where designers have been unable to find a particular fabric to fulfill their needs, and have taken the time to develop a product themselves to realize their vision.

Seymour's personal process starts with information gathering. "If we take on a project in healthcare, we get specialists in that area to speak with us, and if we are talking about sports, we get people from that area. If we are investigating a specific type of workwear, we have to speak with the people who actually work with it. We are looking for everyone, those who work at the company to people who actually wear the garments. This process is very informative and it is also inspirational. I'm inspired a lot by being able to work on all these different products," she says. Gathering information, empathizing with the user, and understanding the market give Seymour and her team multiple points of view into the design of a new product and how it may be used.

Working in context

From there, the Moondial team will start testing individual materials. There is a lot of user analysis, testing interfaces and interactions. They do extensive in-house testing, where people will interact with the garment itself. At the same time they also test the individual materials and technologies to see how far they can go with them. Seymour explains, "We want to see whether it works in context. This is all part of the prototyping phase, and there are usually a lot of prototypes." Once the prototype has been resolved, the Moondial team will oversee the actual production on a consulting basis with each of their clients.

4.79

Sabine Seymour wears the *Orchestra Scarf*.
The project combines sound and textiles
to give the wearer a personalized sonic
experience.

4.80

By manipulating the scarf's various closures,
the wearer can create specific sonic work,
composing a personalized sonic ambience.
The project, commissioned by Moondial, is a
collaboration between Ines Kaag and Desiree
Heiss of BLESS and Ricardo O'Nascimento
of Popkalab.

4.81

The *Recording Shoes* are engineered to
record and play back noises with a built-in
delay; the idea is to create a sensation
of walking when the wearer is actually
standing still.

When describing working on an interdisciplinary team, Seymour says, "Everybody does it differently because everyone comes from a different direction. So if you're an engineer you see it totally differently to if you're a fashion designer—you just think differently." But she acknowledges that the engineering approach has benefits, especially when it comes to hard-core research. Her words of advice are to "collaborate with other people. If you're a designer, go get yourself an engineer. Or if you're an engineer, go get yourself a materials scientist, or collaborate with a company that's creative. This is what I really tell people. You need to be able to collaborate to be able to work with the entire brain."

4.82
A hammock is repurposed to become the *Melodized Pillow Hammock*, an interactive musical performance tool. Rearranging the large pillows will make music, inviting passers-by to experience a unique sonic experience.

Johanna Bloomfield and Ted Southern

Final Frontier Design

Final Frontier Design, founded in 2010, is a high-tech design company based in Brooklyn that aims to craft affordable yet highly capable spacesuits for a burgeoning commercial spaceflight industry. The partners—Ted Southern, a sculptor and costume designer who has built angel wings for Victoria's Secret, and Nikolay Moiseev, a Russian-born mechanical engineer who worked for the Soviet government designing spacesuits—teamed up in 2009 when they won second prize in NASA's Astronaut Glove Competition with their innovative design. Together they have worked with outside consultant Johanna Bloomfield, a fashion designer and technical consultant, on a number of interesting projects, all centered around the single-layer integrated bladder-restraint system that makes their spacesuits unique.

They are currently developing a low-cost commercial inside-the-spacecraft suit called the Intra-Vehicular Activity suit (IVA), which can be pressurized in the event of an emergency. The Final Frontier IVA suit is projected to cost one-fifth of the current NASA IVA suits and will weigh just under 15lb (7kg), half the weight of the NASA suit. On flights with a number of astronauts, that could yield significant fuel savings. Their third-generation prototype, built to NASA flight certification standards, has many new features, including a retractable helmet, improved gloves, and glove disconnects, and is built to withstand greater operating pressure.

The pressure garment

The Final Frontier Design (FFD) team approaches the design process as a problem-solving exercise. They have a series of requirements that are defined by the type of space travel they are designing a suit for, and they build out from there. The foundation of every suit they build is the pressure garment, a full-body airtight suit that provides mechanical counterpressure to assist breathing at altitudes where the air pressure is too low for an unprotected person to be able to survive.

Southern explains, "In the spacesuit world, we are trying to build something that has many different uses: balloon flight, suborbital and orbital flight, and high-altitude jumping. The pressure garment ends up being the universal part of every

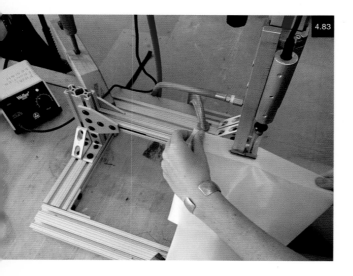

4.83

4.83
Johanna Bloomfield demonstrates the welding machine built at Final Frontier Design to create the single-layer integrated bladder-restraint system used in their uniquely designed spacesuits.

4.84
A finished suit sits on a work table in the studios at Final Frontier Design. On the wall behind are many of the successful spacesuits that the duo has developed, including the work Nikolay Moiseev produced while designing suits for the Soviet space program.

suit and the part that defines the spacesuit. Then we add layers from there. For a high-altitude jump you would need a pressure garment, and some thermal garments, to keep you warm, something to keep you insulated as you jump. But you would not need flame or radiation protection. Our concept is to have a pressure garment then modular-type add-ons that will fit over it according to the mission profile. We spend a lot of time on the pressure garment. It is the hardest part of the suit to design. It is the critical component, it has to have full mobility, it can't leak, and it has to function."

4.85

Various glove prototypes. A successful glove design is very intricate and uses an incredible amount of engineering to function properly in combination with the spacesuits.

4.86

Final Frontier Design logo patch.

4.87

Orange test suit worn in the studio. The team conducts many simple wear simulation tests in the lab to assess the comfort of the astronaut and the functionality of the suit.

Once the problem is outlined, the next step is choosing materials and figuring out how to work with them. Each person on the FFD team plays a specific role in the design process. Moiseev is the engineer, and works on the functional design of the suit; Southern is a people person, and has valuable expertise on working with materials and building prototypes; employee Kari Love is their patternmaker and tailoring expert. Southern explains the dynamic in the studio: "Nikolay and I are partners and we fight like a married couple. He is a very different person to me. He grew up in Soviet Russia, and worked as an engineer designing spacesuits for the government for 20-odd years. He has a different process than I do as an artist. At one point I was having a hard time understanding his motivation behind doing certain things and he said to me [in a Russian

4.85 4.86

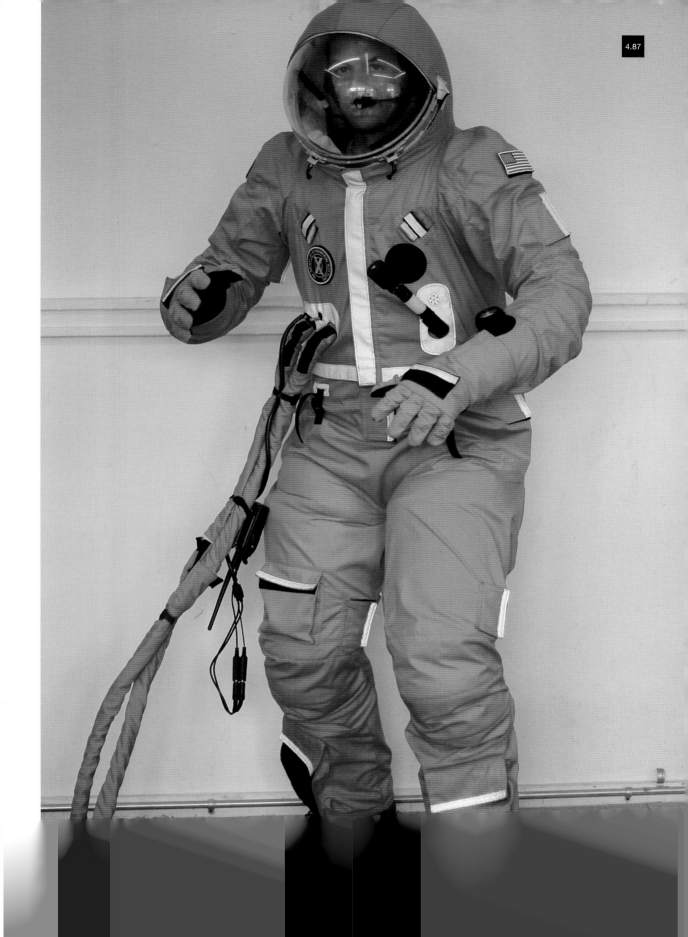

4.88
Nikolay Moiseev (left) and Ted Southern (right)
in the lab.

accent], 'If there was a problem with my design it would be prison for me.' He sums up my life as pleasure-seeking. In the end, our partnership works really well. There are a lot of challenges and communication issues, and we have a lot at stake starting off as entrepreneurs. But I trust Nik, he has a lot of experience."

The OFF Pocket

Southern is also quick to point out the contribution made by Johanna Bloomfield, a multitalented designer who has played an integral part by helping them source key fabrics and finding other collaborators to help them achieve their designs. Bloomfield has also brought to the team an expertise on technical pocket design that came out of a project that she created with the artist Adam Harvey: the OFF Pocket is a privacy protector for your cell phone; once enclosed in the pocket your phone is OFF the network. She has also conducted research into pocket design for the US Army, and it was this that led to consulting with FFD on the design of pockets for the exterior layer of their spacesuit.

Bloomfield's extensive knowledge of technical fabrics and fabric sourcing has led FFD to fabrics that they couldn't have located on their own—specifically a flameproof stretch fabric that they could use in the gloves of their suit. Bloomfield, whose background is in menswear, studied fashion design at the London College of Fashion and worked for Rick Owens, Ralph Lauren, and

Spyder Active Sports before starting her own label and consulting business. She is probably best known for the two projects she did with Adam Harvey, the OFF Pocket and Stealth Wear. She explains, "The idea behind these projects was to create garments that protect the wearer from surveillance without them looking like a cyborg or someone from the future. We wanted to really integrate the protection into street wear with a fashion aesthetic." The OFF Pocket was built in response to the increased tracking of cell phones and is one in a series of pieces, including Stealth Wear, that offer counter-surveillance solutions that allow you to regain control over your privacy.

"Both of these projects worked with metalized fabrics. There was a lot of research and testing involved with the prototyping process. Typically, you wouldn't undertake this in designing a fashion collection. It was a new and challenging experience. This is when I started to focus on sourcing innovative materials and working with FFD. I became their technical materials expert on one of their NASA contracts. One in particular involves radiation for deep space travel," Bloomfield explains.

Both Bloomfield and Southern explain the design process at FFD as more of an engineering-type process. Southern says the studio art training he received at Pratt Institute has served him well. "My professor, Robert Zakarian, always argued that everything has to have a reason. You can't just come up with a color that you like; there is a

right color for everything, there's a right design for everything. There should be some logical reason why you choose something. This is probably more of an engineering attitude than an artist's attitude but I still believe in that."

User testing

Bloomfield adds, "User testing is definitely one of the most important factors that goes into creating innovative products. It is not something that's in a fashion designer's vocabulary. In fashion, you design a product, you put it in a showroom, it goes to market if they buy it. User testing is more related to the world. For example, for Stealth Wear, the OFF Pocket, and definitely for radiation protection, user testing can involve anything from using an infrared camera or a meter that reads radio waves, to getting feedback from actual subjects on the feel of the material, the aesthetic, and how well it works. There is also ASTM (American Society for Testing and Materials) testing, which is the standard testing for the physical properties of materials like breathing

resistance, flame resistance, tensile strength, etc. Standard industry testing that, again, is not something you typically find when you're working on a fashion collection. And then, at the most extreme level, a particle accelerator would be rehired to test high-energy radiation for the radiation coating that FFD is working on."

All of these technical steps in the design process—testing, recording, and measuring the results and comparing them to standards—can take upward of a year. That is why so much of the engineering design approach is spent on materials and construction. When talking about her design process and how she has adapted her traditional fashion design training to accommodate this engineering approach, Bloomfield explains, "My process is really ongoing, there really is no start. But often a new project is inspired with the discovery of a new material. I am often going to smart textile trade shows, performance trade shows, attending talks, giving talks, and taking in information. When I'm working with clients, I help them identify areas of growth where they could integrate a new technology into their line to enhance their product. I am constantly checking periodicals, and searching for every bit of information I can find." Southern agrees, adding that a great source for information is NASA's *Tech Briefs*, a free publication that often has articles on new materials.

Bloomfield continues: "A lot of these new advanced materials do not work on the consumer level yet, although there is beginning to be some crossover. Some of the thermally reflective materials that integrate metals are starting to be used in outerwear."

Working with new materials

As important as the materials is the way they fit together. Both Southern and Bloomfield speak about the difficulty of working with new materials; there is a lot of trial and error involved, even in simply figuring out how to seam them. After a rather lengthy period of trying to sew and seam-seal the urethane-coated nylon fabric they use for their pressure suits, Southern finally discovered that the best method was heat seal. He explains, "Originally I was sewing the fabric and sealing the seams to create an airtight seal. Eventually I called the [fabric] factory and asked them to recommend a sealant. And they said: 'Why are you sealing this? This fabric is heat sealable, you should be ironing this together.' I had no idea." He experimented with an ultrasonic welder, but found that it was best suited to long, straight seams

and was difficult to control for their needs. Finally, they built their own tools—a hand-held device that is essentially a soldering iron with a specially designed tip, and a larger machine that they can manipulate around the tiny seams on the fingers of the spacesuit gloves.

FFD prototype everything in-house. Using their welding machine, and a laser cutter to be able to cut their fabric to within $1/32$in (1mm) tolerance, they create suits that are outperforming standards in testing.

Bloomfield, who consults with other companies, adds that when working with outside factories, sometimes it is "a matter of me, as a designer, understanding their machinery. For example, if they've got an industrial machine with a heavy-duty needle and they're trying to sew a metalized nylon, you might think, 'This is really hard to get the needle through, let's use a bigger needle.' But actually, when you use a bigger needle you're destroying the fabric, creating more leakage points. If you're trying to create an RF [radio frequency electromagnetic radiation] shielding enclosure, you need to communicate with them that they need to change to a smaller needle."

Considering the skills future designers will need, Bloomfield suggests, "Emerging designers should learn to work collaboratively. Design can really help address issues from climate change to the very real prospect that humans will need fully pneumatic wardrobes. They're going to be going to space and back. We're going to require more

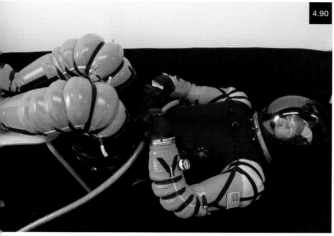

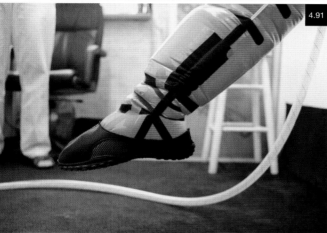

technical garments for that, and a NASA engineer isn't the best person to be designing that garment. I don't want to wear an ASIS [American Society for Industrial Security] suit or a HAZMAT [hazardous materials] suit. I want to wear something that looks like what I'm wearing now, just that on the material level it's more advanced. I think it requires the collaborative spirit and then asking what our environment is going to look like five, ten years from now."

4.90
Simulating an astronaut's seating position at take-off and landing, the user analyzes the suit's fit and function with simple in-lab testing.

4.91
Testing the design of the foot/shoe construction, a user demonstrates the ease and flexibility of the movement while wearing the suit.

4.92
The Final Frontier spacesuit prototype is tested for range of motion and comfort.

Bibliography

Online Articles

Baesgen, H., Schillings, H., Berg, E., (Mar. 20, 1990). "Bioelastic warp-knit and its production". Patent application. Retrieved Dec. 2013. http://www.google.com/patents/US4909049

Boyle, R. (June 2, 2011) "In the future, your clothes will clean the air, generate power and save your life". Retrieved Oct. 2013.

Bradley, R. (Apr. 17, 2012) "Bio-Armor: printing protective plates from patterns in nature". Popular Science. Retrieved July 2013. http://www.popsci.com/technology/article/2012-04/bio-armor

Broudy, B. (Nov. 11, 2011) North Face's ThermoBall wants to revolutionize jacket insulation". Popular Science. Retrieved May 2013. http://www.popsci.com/technology/article/2011-11/north-faces-thermoball-jackets-promise-goldilocks-winter-wa rmth?dom=PSC&loc=recent&lnk=8&con=north-faces-thermoball-wants-to-revolutionize-jacket-insulation

Broudy, B. (Aug. 1, 2012) "The first short that lowers your body temperature". Popular Science. Retrieved July 2013. http://www.popsci.com/technology/article/2012-07/first-shirt-lower-body-temperature

Cartlidge, E. (May 10, 2011) "Translucent curtains soak up sound". IOP Institute of Physics. Retrieved Nov. 2013 http://physicsworld.com/cws/article/news/2011/may/10/translucent-curtains-soak-up-sound

Cochlin, D. (Sept. 4, 2012) "'Magic carpet' could help prevent falls". University of Manchester. Retrieved Feb. 2013. http://www.manchester.ac.uk/discover/news/article/?id=8648

Collette, M. (Jan. 4, 2012) "With tentacles in many disciplines, capstone team merges engineering, design." News at Northeastern. Retrieved Jan. 2014. http://www.northeastern.edu/news/2012/01/squid/

Coxworth, B. (May 2, 2012) "Squid-inspired tech could lead to color-changing smart materials". Gizmag. Retrieved Sept. 2013. http://www.gizmag.com/squid-inspired-color-changing-clothes/22383/

Crane, L. (June 21, 2013) "Under Armour 39 review". Digital Trends. Retrieved Dec. 2013. http://www.digitaltrends.com/fitness-apparel-reviews/under-armour39-review/#!bCcS1Y

Diep, F. (Feb. 27, 2013) "Insanely rubbery battery stretches to 4 times its length". Popular Science. Retrieved July 2013. http://www.popsci.com/technology/article/2013-02/rubbery-battery-stretches-300-percent

Empson, E. (June 30, 2012) "With tech from space, Ministry of Supply is building the next generation of dress shirts". Tech Crunch. Retrieved Sept. 2013. http://techcrunch.com/2012/06/30/ministry-of-supply/

Fang, J. (Sept. 25, 2013) "The smart textiles of tomorrow". Fashiontech. Retrieved Mar. 2014. http://fashiontech.wordpress.com/2013/09/25/7630/

Ferro, S. (Mar. 5, 2013) "How winter woes inspired a nanotech fix for everything from cold necks to knee pain". Popular Science. Retrieved June 2013. http://www.popsci.com/technology/article/2013-03/tech-transfer-winter-woes-nanotech-cold-necks-knee-pain?dom=PSC&loc=recent&lnk=1&con=how-winter-woes-inspired-a-nanotech-fix-for-everything-from-cold-necks-to-knee-pain

Fox, S. (Apr. 25, 2012) "2012 Military wishlist features smart wound-diagnosing uniforms and dogfighting drones". Popular Science. Retrieved Aug. 2013. http://www.popsci.com/technology/article/2012-04/2012-military-wishlist-features-smart-wound-diagnosing-uniforms-and-dogfighting-drones?dom=PSC&loc=recent&lnk=6&con=2012-military-wishlist-features-smart-wounddiagnosing-uniforms-and-dogfighting-drones

Jirousek, C. (1995) "Creativity and the design process". Art Design and Visual Thinking. Retrieved Jan. 2013. http://char.txa.cornell.edu/language/creative.htm

Klausner, A. (?). "Slipping into smart fabrics". Core 77. Retrieved Sept. 2013. http://www.core77.com/materials/art_smartfab.asp

Lecher, C. (June 13, 2013) "'NeuroKnitting' turns brain scans Into personalized scarves". Popular Science. Retrieved May 2013. http://www.popsci.com/technology/article/2013-06/neuroknitting-turns-brain-scans-personalized-scarves?dom=PSC&loc=recent&lnk=1&con=neuroknitting-turns-brain-scans-into-personalized-scarves

Meinhold, B. (Sept. 30, 2011) "Under Armour's biometric compression shirt tracks, broadcasts athletic performance (video)". Ecouterre. Retrieved Nov. 2013. http://www.ecouterre.com/under-armours-biometric-compression-shirt-tracks-broadcasts-athletic-performance-video/

Nosowitz, D. (Apr. 7, 2011) "New superhydrophobic fabric blocks both water and UV rays". Popular Science. Retrieved June 2013. http://www.popsci.com/technology/article/2011-04/new-superhydrophobic-fabric-blocks-both-water-and-uv-rays

Rossiter, J., Yap, B., Conn, A., (May 2, 2012). "Squid and zebrafish cells inspire camouflage smart materials". IOP Institute of Physics. Retrieved Nov. 2013. https://www.iop.org/news/12/may/page_55183.html

Syuzi, (Dec. 16, 2009) "CO2 dress – a beautiful pollution-sensing dress". Fashioning Tech. Retrieved Nov. 2013. http://fashioningtech.com/profiles/blogs/c02-dress-a-beautiful

Templeton, G. (June 30, 2013) "New smart fiber changes color when stretched". Geek.com. Retrieved Aug. 2013. http://www.geek.com/science/new-smart-fiber-changes-color-when-stretched-1537741/

Yu, B. (?). Textile damage. *International Fabricare Institute Bulletin, #629, 5/91* http://70.88.161.72/ifi/BULLETIN/TOI/Toi629.pdf

Printed Articles

Chuang, M., Windmiller, J. et al. "Textile-based Electrochemical Sensing: Effect of Fabric Substrate and Detection of Nitroaromatic Explosives". Electroanalysis ISEAC 2012, Volume 22, pages 2511–2518, Nov. 2010.

Hamedi, M., Herlogsson, L., Crispin, X., Marcilla, R. et al. "Electronic Textiles: Fiber-embedded Electrolyte-gated Field-effect Transistors for e-Textiles". Wiley Online Library. John Wiley & Sons, Inc., 22 Jan. 2009.

Hamedi, M., Forchheimer, R., Inganas, O., "Towards Woven Logic from Organic Electronic Fibres". Nature Materials. Nature Publishing Group, 4 Apr. 2007.

Lee, M., Eckert, R., Forberich, K., Dennler, G., et al., "Solar Power Wires Based on Organic Photovoltaic Materials". Science. American Association for the Advancement of Science, 12 Mar. 2009.

Malzahn, K. Windmiller, JR, et al. "Wearable Electrochemical Sensors for in situ Analysis in Marine Environments". Analyst. 2011 July 21;136(14):2912-7.

Post, R., Orth, M., Russo, P., and Gershenfeld, N. "E-broidery: Design and Fabrication of Textile-based Computing." IBM Systems Journal 39, 3-4 (2000), 840–860.

Windmiller, J. and Wang, J. "Wearable Electrochemical Sensors and Biosensors: a Review". Electroanalysis ISEAC 2012, Volume 25, Issue 1, pages 29-46.

Yang, Y., Chuand, M., Lou, S., Wang, J. "Thick-film Textile-based Amperometric Sensors and Biosensors". Analyst 2010.

Books

Braddock Clark, S., O'Mahony, M. *SportsTech: revolutionary fabrics, fashion and design.* New York: Thames & Hudson, 2002.

Braddock Clark, S., O'Mahony, M. *Techno textiles 2.* New York: Thames & Hudson, 2006.

Damon, A., H. Stoudt and R. McFarland, *The human body in equipment design.* Cambridge, Ma,: Harvard University Press, 1966.

Hatch, K., *Textile science.* Minneapolis: West Publishing Company, 1993.

Hudson, P., Clapp, A., Kness, D., *Joseph's introductory textile science, sixth edition.* New York: Harcourt Brace, 1993.

Jones, C., et al., *Sensorium embodied experience, technology and contemporary art.* Massachusetts: The MIT Press, 2006.

Joseph, M., *Introductory textile science, sixth edition.* New York: Holt, Rinehart and Winston, 1992.

McQuaid, M., et al. *Extreme textiles.* New York: Princeton Architectural Press, 2005.

Moritz, E., et al., *The engineering of sport, volume 3,* Germany: Springer Science + Business Media, 2010.

Quinn, B., *Techno fashion.* London, New York: Berg 2002.

Quinn, B., *Textile futures.* London, New York: Berg 2010.

Quinn, B., *Textile visionaries.* London: Laurence King Publishing, 2013.

Raheel, M., *Protective clothing systems and materials.* New York: Marcel Dekker, Inc. 1994.

Renbourn, E., *Physiology and hygiene of materials and clothing.* Watford, UK: Merrow Series, 2004.

Renbourn, E. T., and W. T. Rees, *Materials and clothing in health and disease.* London: H. K. Lewis, 1998.

Salazar, L., *Fashion v sport.* London: V&A Publishing, 2008.

Seymour, S., *Fashionable technology.* New York: Springer Wien New York, 2009.

Stewart, R., *Higher further, faster...is technology improving sport?* UK: John Wiley & Sons, Inc., 2008.

Index

Page numbers in *italics* refer to illustration captions

Photo credits

11: the Holy dress was made by: Melissa Coleman, Leonie Smelt and Joachim Rotteveel, photography: Sanja Marušić; 14: © Kim Kyung Hoon/Reuters/Corbis; 15t: Flori Kryethi EOTEVI; 15b: Living Wall, courtesy Leah Buechley; 16: Florian Kräutli; 17l, c: Pierre Verdy/AFP/Getty Images; 17r, 115, 116, 117: polychromelab; 18: courtesy Mercedes-Benz UK Limited; 19, 82, 83, 84, 85, 91, 101r: courtesy GZE; 20l: designer: Angella Mackey, photographer: Henrik Bengtsson, model: Jenny Andersson; 20r: designed by Diffus Design Aps by Hanne-Louise Johannesen and Michel Guglielmi, developed by Diffus Design Aps together with Inntex, Italy, embroidery by Forster Rohner AG and their revolutionary embroidery® technology, image © Diffus Design Aps; 21: Neuroknitting, by Varvara Guljajeva, Mar Canet and Sebastian Mealla Cincuegrani (2013), model: Sebastian Mealla Cincuegrani; 22, 23: the 'Herself' dress is the world's first prototype air purifying dress, the piece is a key output of The Catalytic Clothing Project, in collaboration with Professor Tony Ryan PVC, Faculty of Science at Sheffield University, the textiles were created as part of a collaboration with Trish Belford of Tactility Factory and the University of Ulster in 2010, image by DED Design; 24l: © Steven Tee/LAT Photographic; 24r: © Sutton Images/Corbis; 25l: SSG Coltin Heller/DVIDS; 25r: inventor, science and engineering: Professor Dava Newman, MIT, design: Guillermo Trotti, A.I.A., Trotti and Associates, Inc.,Cambridge, MA, fabrication: Dainese, Vicenza, Italy, photography: Douglas Sonders; 26, 27: Pankaj & Nidhi; 29: Novanex Interactive LED dress, 2011, © Sonago for Novanex; 30, 69: Francis Bitonti Studio Inc.; 31: photo: Greg Kessler; 32, 33: photo: Martin Noboa; 36: Adam Paxson, Kyle Hounsell, & Jim Bales; 37: courtesy Spidi Sport SRL; 38l: marekuliasz/Shutterstock.com; 38r: Rex Features; 39: © Outlast Technologies LLC; 40: Schoeller Technologies AG; 41l: courtesy Speedo and ANSYS, Inc.; 41r: dpa picture alliance archive/Alamy; 42l: Hannah Peters/Getty Images; 42r: Peter Parks/AFP/Getty Images; 44l: Sail Racing Int/Jan Söderström; 44r: Jeff Harris/Gallery Stock; 45tl: project: Greg Lynn FORM, North Sails, Swarovski, image courtesy Greg Lynn FORM; 45tr: Paul Todd/Gallo Images/Getty Images; 45b: Don Emmert/AFP/Getty Images; 46l, 56l: courtesy of Less EMF Inc.; 46r: potowizard/Shutterstock.com; 47: photos taken by Dr Patrick Hook, inventor of the helical-auxetic system and Managing Director, Auxetix Ltd.; 48: NASA; 49t: Shane Gross/Shutterstock.com; 49bl: Eye of Science/Science Photo Library; 49br: Pascal Goetgheluck/Science Photo Library; 50t: designed by Noah Waxman in Brooklyn, NY, photo: Owen Bruce; 50bl: © Science Picture Co./Corbis; 50br: image provided by Jianfeng Zang and Zuanhe Zhao, Duke University; 51: photography: Özgür Albayrak, © Max Schäth, e-motion project, Berlin University of the Arts, Prof.

Valeska Schmidt-Thomsen, 2009; 52l: Adam Paxson; 52r: www.skin-care-forum.basf.com, BASF Personal Care and Nutrition GmbH; 53: © Habu Textiles, photography by Vanessa Yap-Einbund; 54, 55, 56, 59, 88: Forster Rohner Textile Innovations; 57l: courtesy Parabeam B.V.; 57c: photo: Lorenzo Marchesi; 57r: Someya-Sekitani Group, University of Tokyo; 58tl: designer: Pauline van Dongen, this project was developed in collaboration with Christiaan Holland and Gert Jan Jongerden, photography: Mike Nicolaassen, hair & make-up: Angelique Stapelbroek, model: Julia J. at Fresh Model Management; 58tr: courtesy the artist and Klaus von Nichtssagend Gallery; 58b: courtesy Jin-Woo Han, Ames Research Centre, NASA; 60, 61: Zane Berzina and Jackson Tan; 62, 74l: courtesy Loop.pH; 63l: photo: Eva Stööp; 63r: courtesy Inventables Inc.; 64l: Jonathan Rossiter & Andrew Conn, Univ. Bristol, UK; 64c: H.S. Photos/Alamy; 64r: US Department of Energy/Science Photo Library; 65tl, tr: The North Face®; 65b: Col D'Anterne, France, photo: Damiano Levati; 66: Demon United, 2014 www.demonsnow.com; 67: photo courtesy of Dow Corning; 68: Neri Oxman, Architect and Designer, in collaboration with W. Craig Carter (MIT), 2012, Digital Materials, Museum of Science, Boston; 70t: made by Alice Nasto, in the MIT Media Lab course "New Textiles" taught by Leah Buechley; 70c, b: courtesy Earl Stewart; 71: Strawberry Noir, part of the Biolace series © Carole Collet 2012; 74r: created by Aaron Sherwood & Michael Allison, photo by Yang Jiang; 76: courtesy Jennifer Darmour; 78, 80: www.Philips.com; 79t: Kollektion Silent Space © Annette Douglas Textiles; 79b: courtesy Roel Vertegaal; 81l: project: concept plane AIRBUS A350, designers flooring: Erik Mantel & Yvonne Laurysen, company: LAMA concept, interior design of Airbus: Priestman Goode & Airbus Design Team, year: 2006, courtesy image: AIRBUS France; 81r: designers: Erik Mantel & Yvonne Laurysen, company: LAMA concept, year: 2006, courtesy image: LAMA concept; 86: CITA - Centre for IT and Architecture, Copenhagen, by Mette Ramsgard Thomsen and Karin Bech; 87: creator: Dr Jenny Tillotson, Textile Futures Research Centre, University of the Arts, London, photographer: Simon Barber; 89: courtesy Columbia Sportswear; 90: warmX GmbH German, Mr. Christoph Mueller; 92l: Design Mithril Kevlar Jacket: Peter Askulv, Klättermusen; 92r: photo: Helly Hansen AS; 93: photography: Jonathan Shaun/Makers & Riders, designer/maker of pant: Jonathan Shaun/Makers & Riders, design: 3-Season Weatherproof Suprema Jean, textile: Polartec NeoShell Waterproof, handmade in Chicago, IL; 94: Elisabeth Grebe, Linz; Copyright: Utope 2014, www.utope.eu; 95: Joe Robbins/Getty Images; 96, 105: courtesy MC10; 97: courtesy DuPont; 98: courtesy Garrison Bespoke; 99: Hövding (photographer: Jonas Ingerstedt);

Acknowledgements

In the creation of this book, many designers, artists, and scientists took time out of their incredibly busy schedules for my interviews and replied to my inquiries with amazing speed and willingness to share their thoughts, work, and enthusiasm. I am also deeply grateful to all the amazing individuals and companies that provided me with images of their work. The overwhelming response to my queries both humbled and inspired me. Thank you to all, this book would not have come together without all of you.

A huge thank you to my publisher, Laurence King Publishing of London and my editor, Peter Jones, for his incredible insight, invaluable criticism, and patience during the entire creation process.

Finally, I am indebted to my husband, Daniel, and my sons, Dakota and Dylan, who provided support and encouragement at every stage of the book. This book is dedicated to them.